T0215797

Cloth Simulation for Computer Graphics

Synthesis Lectures on Visual Computing: Computer Graphics, Animation, Computational Photography, and Imaging

Editor
Brian A. Barsky, *University of California, Berkeley*

This series presents lectures on research and development in visual computing for an audience of professional developers, researchers, and advanced students. Topics of interest include computational photography, animation, visualization, special effects, game design, image techniques, computational geometry, modeling, rendering, and others of interest to the visual computing system developer or researcher.

Cloth Simulation for Computer Graphics
Tuur Stuyck
2018

Design, Representations, and Processing for Additive Manufacturing
Marco Attene, Marco Livesu, Sylvain Lefebvre, Thomas Funkhouser, Szymon Rusinkiewicz, Stefano Ellero, Jonàs Martínex, and Amit Haim Bermano
2018

Virtual Material Acquisition and Representation for Computer Graphics
Dar'ya Guarnera and Giuseppe Claudio Guarnera
2018

Stochastic Partial Differential Equations for Computer Vision with Uncertain Data
Tobias Preusser, Robert M. Kirby, and Torben Pätz
2017

iii

An Introduction to Laplacian Spectral Distances and Kernels: Theory, Computation, and Applications
Giuseppe Patanè
2017

Mathematical Basics of Motion and Deformation in Computer Graphics, Second Edition
Ken Anjyo and Hiroyuki Ochiai
2017

Digital Heritage Reconstruction Using Super-resolution and Inpainting
Milind G. Padalkar, Manjunath V. Joshi, and Nilay L. Khatri
2016

Geometric Continuity of Curves and Surfaces
Przemyslaw Kiciak
2016

Heterogeneous Spatial Data: Fusion, Modeling, and Analysis for GIS Applications
Giuseppe Patanè and Michela Spagnuolo
2016

Geometric and Discrete Path Planning for Interactive Virtual Worlds
Marcelo Kallmann and Mubbasir Kapadia
2016

An Introduction to Verification of Visualization Techniques
Tiago Etiene, Robert M. Kirby, and Cláudio T. Silva
2015

Virtual Crowds: Steps Toward Behavioral Realism
Mubbasir Kapadia, Nuria Pelechano, Jan Allbeck, and Norm Badler
2015

Finite Element Method Simulation of 3D Deformable Solids
Eftychios Sifakis and Jernej Barbič
2015

Efficient Quadrature Rules for Illumination Integrals: From Quasi Monte Carlo to Bayesian Monte Carlo
Ricardo Marques, Christian Bouville, Luís Paulo Santos, and Kadi Bouatouch
2015

Numerical Methods for Linear Complementarity Problems in Physics-Based Animation
Sarah Niebe and Kenny Erleben
2015

Mathematical Basics of Motion and Deformation in Computer Graphics
Ken Anjyo and Hiroyuki Ochiai
2014

Mathematical Tools for Shape Analysis and Description
Silvia Biasotti, Bianca Falcidieno, Daniela Giorgi, and Michela Spagnuolo
2014

Information Theory Tools for Image Processing
Miquel Feixas, Anton Bardera, Jaume Rigau, Qing Xu, and Mateu Sbert
2014

Gazing at Games: An Introduction to Eye Tracking Control
Veronica Sundstedt
2012

Rethinking Quaternions
Ron Goldman
2010

Information Theory Tools for Computer Graphics
Mateu Sbert, Miquel Feixas, Jaume Rigau, Miguel Chover, and Ivan Viola
2009

Introductory Tiling Theory for Computer Graphics
Craig S.Kaplan
2009

Practical Global Illumination with Irradiance Caching
Jaroslav Krivanek and Pascal Gautron
2009

Wang Tiles in Computer Graphics
Ares Lagae
2009

Virtual Crowds: Methods, Simulation, and Control
Nuria Pelechano, Jan M. Allbeck, and Norman I. Badler
2008

Interactive Shape Design
Marie-Paule Cani, Takeo Igarashi, and Geoff Wyvill
2008

Real-Time Massive Model Rendering
Sung-eui Yoon, Enrico Gobbetti, David Kasik, and Dinesh Manocha
2008

High Dynamic Range Video
Karol Myszkowski, Rafal Mantiuk, and Grzegorz Krawczyk
2008

GPU-Based Techniques for Global Illumination Effects
László Szirmay-Kalos, László Szécsi, and Mateu Sbert
2008

High Dynamic Range Image Reconstruction
Asla M. Sá, Paulo Cezar Carvalho, and Luiz Velho
2008

High Fidelity Haptic Rendering
Miguel A. Otaduy and Ming C. Lin
2006

A Blossoming Development of Splines
Stephen Mann
2006

© Springer Nature Switzerland AG 2022
Reprint of original edition © Morgan & Claypool 2018

All rights reserved. No part of this publication may be reproduced, stored in a retrieval system, or transmitted in any form or by any means—electronic, mechanical, photocopy, recording, or any other except for brief quotations in printed reviews, without the prior permission of the publisher.

Cloth Simulation for Computer Graphics
Tuur Stuyck

ISBN: 978-3-031-01469-7   paperback
ISBN: 978-3-031-02597-6   ebook
ISBN: 978-3-031-00348-6   hardcover

DOI: 10.1007/978-3-031-01769-7

A Publication in the Springer series
Synthesis Lectures on Visual Computing: Computer Graphics, Animation,
Computational Photography, and Imaging

Lecture #32
Series Editor: Brian A. Barsky, *University of California, Berkeley*
Series ISSN
Print 2469-4215   Electronic 2469-4223

Cloth Simulation for Computer Graphics

Tuur Stuyck

SYNTHESIS LECTURES ON VISUAL COMPUTING: COMPUTER GRAPHICS, ANIMATION, COMPUTATIONAL PHOTOGRAPHY, AND IMAGING #32

ABSTRACT

Physics-based animation is commonplace in animated feature films and even special effects for live-action movies. Think about a recent movie and there will be some sort of special effects such as explosions or virtual worlds. Cloth simulation is no different and is ubiquitous because most virtual characters (hopefully!) wear some sort of clothing.

The focus of this book is physics-based cloth simulation. We start by providing background information and discuss a range of applications. This book provides explanations of multiple cloth simulation techniques. More specifically, we start with the most simple explicitly integrated mass-spring model and gradually work our way up to more complex and commonly used implicitly integrated continuum techniques in state-of-the-art implementations. We give an intuitive explanation of the techniques and give additional information on how to efficiently implement them on a computer.

This book discusses explicit and implicit integration schemes for cloth simulation modeled with mass-spring systems. In addition to this simple model, we explain the more advanced continuum-inspired cloth model introduced in the seminal work of Baraff and Witkin [1998]. This method is commonly used in industry.

We also explain recent work by Liu et al. [2013] that provides a technique to obtain fast simulations. In addition to these simulation approaches, we discuss how cloth simulations can be art directed for stylized animations based on the work of Wojtan et al. [2006]. Controllability is an essential component of a feature animation film production pipeline. We conclude by pointing the reader to more advanced techniques.

KEYWORDS

physics-based simulation, cloth simulation, computer graphics, explicit integration, implicit integration, adjoint optimization

Contents

Preface xiii

Acknowledgments xv

1   Introduction 1
   1.1   Physics-Based Animation 1
   1.2   Applications of Cloth Simulation 2
       1.2.1   Offline Simulations 2
       1.2.2   Real-Time Simulation 3
   1.3   Cloth Simulation Pipeline in Animation 3
       1.3.1   Research 3
       1.3.2   Software Development 4
       1.3.3   Simulation in Production 4
   1.4   History of Cloth Simulation 4
   1.5   Overview of This Book 5
   1.6   Intended Audience 6
   1.7   Getting Started 6

2   Cloth Representation 9
   2.1   Triangles 9
   2.2   Particles 11
   2.3   Forces 12
       2.3.1   Frames and Steps 12

3   Explicit Integration 13
   3.1   Introduction 13
   3.2   Explicit Integration 13
   3.3   Stability Analysis 15
       3.3.1   Test Equation 16
       3.3.2   Explicit Euler Analysis 17
   3.4   Adaptive Time Stepping 19
   3.5   Conclusion 19

4 Mass-Spring Models **21**

4.1 Introduction 21

4.2 Computing Masses 21

4.3 Computing Forces 22

    4.3.1 Energy Minimization 23

    4.3.2 Spring Potential Energy and Force 24

    4.3.3 Spring Damping Force 26

4.4 Putting It All Together 27

4.5 Tearable Cloth 27

4.6 Other Mass-Spring Applications 28

    4.6.1 Hair Simulation 28

    4.6.2 Soft Body Dynamics 28

4.7 Conclusion 29

5 Implicit Integration **31**

5.1 Introduction 31

5.2 Backward Euler 31

    5.2.1 Linearization 32

5.3 Stability Analysis 34

5.4 Spring Forces and Their Derivatives 35

5.5 Block Compressed Row Storage 38

    5.5.1 Matrix-Vector Multiplication 40

5.6 Adding Velocity Constraints 40

5.7 Solving the Linear System 41

    5.7.1 Preconditioning 42

5.8 Position Alterations 42

5.9 A Quick Note on Stability 44

5.10 Alternative Integration Schemes 44

5.11 Conclusion 45

6 Simulation as an Optimization Problem **47**

6.1 Introduction 47

6.2 Notation 48

6.3 Reformulating the Problem 48

6.4 Solving the Nonlinear Actuations 49

6.5 Local-Global Alternation Problem Formulation 51

6.6   Solving Time Integration Using Local-Global Alternation 54
      6.6.1 Local Step 54
      6.6.2 Global Step 54
6.7   Conclusion 56

7   Continuum Approach to Cloth **57**
7.1   Introduction 57
7.2   Cloth Rest Shape 57
7.3   Computing Forces and their Derivatives 58
      7.3.1 Damping Forces 61
7.4   Stretch Forces 63
7.5   Shear Forces 69
7.6   Bend Forces 71
7.7   Conclusion 73

8   Controlling Cloth Simulations **75**
8.1   Introduction 75
8.2   Control Problem Formulation 76
      8.2.1 The Goal Function 76
      8.2.2 Tuning the Goal Function 77
      8.2.3 Minimizing the Goal Function 78
8.3   Adjoint State Computation 79
8.4   Updating Control Forces 81
8.5   Creating Keyframes 83
8.6   Conclusion 84

9   Collision Detection and Response **85**
9.1   Introduction 85
9.2   Collision Detection 86
      9.2.1 Bounding Volume Hierarchies 87
      9.2.2 Basic Primitive Tests 88
9.3   Collision Response 88
      9.3.1 Cloth-Cloth Collision Response 88
      9.3.2 Object-Cloth Collision Response 89
9.4   Discussion 89
9.5   Further Reading 90
9.6   Conclusion 90

10   What's Next **91**

   10.1   Real-Time Applications 91

   10.2   Subspace Cloth Simulation 92

   10.3   Alternative Cloth Models 92

   10.4   Art Directing Cloth 93

   10.5   Cloth Rendering 93

11   Conclusions **95**

A   Vector Calculus **97**

   Bibliography **99**

   Author's Biography **105**

Preface

This book has grown from a desire to make cloth simulation more accessible to people new to the field. It is the hope that this book serves as a good practical guide to bring you up to speed to allow you to implement your own cloth simulator and produce visually pleasing results.

The literature on cloth simulation is very vast and new work is published every year. The intention of this book is not to cover all the topics but rather that this tutorial will provide a solid understanding of the basics so that you will more easily understand technical papers that build upon these foundations.

Tuur Stuyck
July 2018

Acknowledgments

I would like to thank Amanda Ha for proofreading this book and for supporting me throughout the process. Emilee Chen, thank you for motivating me to start working on this manuscript. I thank Toon Stuyck, Dries Verhees, Nathan Waters, and Erik Englesson for providing me with valuable feedback during the writing of this book.

Special thanks to Donald House, David Eberle, Witawat Rungjiratananon, and Armin Samii for their insightful comments.

This book wouldn't have come to fruition without the support of my family and the people at Morgan & Claypool publishing. Thank you!

David Eberle and Kurt Fleischer, thank you for being inspiring mentors and great friends.

Tuur Stuyck
July 2018

CHAPTER 1

Introduction

Making a feature-length computer-animated movie costs millions of U.S. Dollars and takes several years of planning, script writing, visual development, and eventually modeling and animation on a computer. The computer graphics community is pushed forward by solving the challenges artists are faced with during the development of new movies. One of these research areas is physics-based animation where engineers and researchers use physics and math to make beautiful animations of natural phenomena.

1.1 PHYSICS-BASED ANIMATION

The use of physically based simulations is ubiquitous in games and special effects for movies. As an animator, you really don't want to be tasked with animating water or cloth by hand since these materials need to follow strict physical laws in order to be plausible and believable to the audience. Every one of us knows how water is supposed to behave so it is very difficult and time consuming to recreate this by hand.

This is where simulation shines. We can model the real world in the computer and compute how the materials would behave under the influence of the environment. A few examples of simulations are fluid simulations that are used for modeling flowing rivers, explosions, and smoke. Other applications are the simulation of rigid body interactions such as the destruction and collapsing of a building. Also, soft body simulations such as flesh and muscle simulations are used for virtual surgery. Of course, there is also cloth simulation that will allow artists to obtain detailed and natural looking geometry that reacts to the movement of the character and wind forces.

Highly believable simulated motions are typically generated by numerical algorithms that evolve discrete mathematical models over time. The model describes how the material should move, taking into account the material properties, boundaries, external forces, and collision objects in the scene. In computer graphics, we are mostly concerned with the look of the final result and physical accuracy is by no means our main goal. This is in stark contrast to the engineering community. For their purposes, physical accuracy is a top priority in order to be able to run simulations that are helpful in modeling and predicting real-world-scenarios. Obviously, physical accuracy helps achieve these visually pleasing and physically plausible goals for applications in computer graphics.

1.2 APPLICATIONS OF CLOTH SIMULATION

Before we overload you with mathematical expressions and derivations, let's get you excited by talking about common applications of cloth simulation. The applications can typically be categorized in one of the following two categories.

- **Offline simulations** are computed, tweaked, and post processed before being rendered on screen. The artist has time to run multiple simulations with different settings in order to find the desired results. These methods typically target high believability and controllability.

- **Real-time simulations** involve computing the simulation dynamics at runtime. This will allow the simulation to interactively react to user input and changes in the virtual environment. This type of simulations have very limited computation time available to them and are commonly implemented on GPU hardware. Real-time simulation algorithms are required to be fast and stable.

Specific examples of both categories are given in the following subsections.

1.2.1 OFFLINE SIMULATIONS

The most obvious applications are the use in special effects, digital doubles, computer animation, and virtual prototyping. The special effects industry has advanced so much that, instead of hiring a stunt double, it is sometimes easier to just digitally recreate the actor. This requires that we can also accurately model their clothing and the way the cloth behaves. That way, a smooth transition can be made from the real actor to the digital double, leaving the viewer none the wiser on how they performed the actual stunt. Spoiler: it's all computers and the amazing craftsmanship and dedication of animators and technical directors.

Computer animation is very similar to special effects, although the focus is often a little different. Special effects want to stay close to reality in order to truthfully recreate actors. Computer animation often involves virtual characters created by the director and their highly talented development team. Their focus is artistic expression. Directors are very concerned with being able to convey a very stylized style in order to tell the story the best way they can. The focus in computer animation is thus mostly controllability and art directability.

As a last example, fashion designers can use virtual cloth models to find the right 2D patterns that make up garments. A computer implementation of the cloth dynamics allows them to quickly iterate on designs and visualize how the garment will drape and where folds and wrinkles will be created naturally due to the material and sewing patterns. Virtual prototyping allows them to save on material and fabrication costs and accelerates the design process.

1.2.2   REAL-TIME SIMULATION

The most obvious candidate for real time simulations is of course computer games. Expectations about the visual gorgeousness for AAA games keep rising given the increasing computational power available in desktop machines and new generations of game consoles. This puts a lot of pressure on game developers to produce extremely efficient implementations that make the most of the available hardware.

Interactive physical simulations contribute to obtaining an immersive experience for the player. For current generation consoles, 30 frames per second is the standard. For applications such as virtual reality or pc games, 60 frames per second is the norm.

A screen refresh rate of 60 times per second means that we only have 16 milliseconds of time to compute a new frame. That's not a lot at all! Especially considering that this is the total frame time. In this limited time, we need to compute the rendering, human-computer interactions, networking, artificial intelligence of the digital agents, and, of course, the physics simulation. Many commercial games also use the cloth simulation pipeline for simulating hair. This makes its efficiency even more important.

Other applications are virtual reality where the user is fully immersed in the virtual environment or augmented reality where a virtual layer is overlayed on the real world. One other upcoming application is virtual fitting rooms. An incredible amount of clothing is purchased every day from online retailers. A large number of these items are returned because they're not the correct size. Imagine having your own digital double that would allow you to virtually shop for clothing and fit them, finding the right size, all without having to leave the house!

1.3   CLOTH SIMULATION PIPELINE IN ANIMATION

The amazing simulations that are brought to your screen were touched by the hands of many people. In this section, we'll give a quick overview of the different steps involved in the process.

1.3.1   RESEARCH

It all starts with research scientists, academics, and engineers developing new simulation models. These are frequently published in computer graphics journals and presented at conferences such as *SIGGRAPH*, *SIGGRAPH ASIA*, and *EUROGRAPHICS*, among others. Every newly published paper presents some significant improvement over previous methods. Research typically happens in academic institutes such as universities or in research labs in industry.

Academic research is essential but papers often only show results on sandboxed examples that aren't necessarily as complex as real production scenes. This isn't a bad thing since the papers needs to show validation of the method which is often a very specific aspect of cloth simulation.

1.3.2   SOFTWARE DEVELOPMENT

The next step in the process is making the scientific work more robust for use in production or commercial software. This type of work is frequently described with professional titles such as research engineer or simulation developer. The job often involves integrating new methods into existing legacy codebases that many people use and depend on. As well as improving robustness by handling numerous edge-cases and unforeseen use-cases.

Academic work often makes certain assumptions such as intersection-free animation or manifold meshes. These assumptions are definitely not guaranteed in production and violations need to be resolved in the codebase. Additionally, new techniques often require user input such that custom interfaces need to be implemented to expose these features to the user. The software team should work closely with their users, the artists. They are additionally tasked with implementing feature requests and fixing bugs and improving the overall pipeline.

Another big focus of engineers is to obtain the best efficiency possible. Artists use the software on a daily basis and require fast turnaround times so that they can quickly iterate on their work. Having idle artists waiting for simulations to finish is frustrating for all parties involved. Not just that, simulations consume computing resources that are often shared with other departments such as the rendering department. You won't be making any friends with other departments when your simulations are hogging all the machines on the render farm.

1.3.3   SIMULATION IN PRODUCTION

Artists working in the simulation department are called simulation artists or simulation technical directors. They are tasked with creating the simulations that are shown on screen. They will need to use their own judgement and take feedback from the simulation leads and directors to create simulations that follow the creative vision of the director. This is a very labor-intensive and deadline-sensitive job that requires both artistic skill and technical knowledge.

The tailoring team will model garments and set-up cloth parameters based on concept art. This requires expertise in tailoring techniques. These garments are then set-up in shots where the technical directors tweak parameters and add forces such as wind and numerous other constraints to achieve the desired look. For art-directed films, simulation rarely looks the way they want out-of-the-box. Creating the desired look and feel often involves trial-and-error, running multiple simulations to visualize the effects of changing simulation settings.

Once the motion of the simulations are finalized, other teams will take care of shading and rendering.

1.4   HISTORY OF CLOTH SIMULATION

Cloth simulation has been an active field of research for decades. It has been extensively researched by material and fabric scientists. The first advances in computer graphics were made in the eighties by Weil [1986]. He developed the first computer graphics models for mimicking the drape of fabrics that are held at constraint points. The method was purely geometry based and produced results that look draped and wrinkly, just like cloth. The model didn't incorporate cloth movement yet but opened up cloth research to the graphics community.

Around the same time, researchers became interested in more physically based techniques to model the cloth behavior. Early work can be found in the thesis of Feynman [1986]. He modeled cloth as an elastic sheet using a continuum representation. Continuum mechanics models physical properties and movement using a continuous mass representation rather than discrete particles. A more general method for elastic modeling with cloth applications was later developed by Terzopoulos et al. [1987]. These techniques rely on the assumption that the material in question can accurately be modeled using continuum mechanics. This is a reasonable assumption for plastics or rubber for which structure is only visible at a microscopic or molecular level. On the other hand, cloth is a woven material where the structure is visible to the naked eye. Follow-up research in the nineties was aimed at finding a good continuum representation, specifically for cloth. The results can be found in the work of Carignan et al. [1992].

In contrast to continuum methods, Haumann [1987] and Breen et al. [1992] worked on particle-based simulations of clothing. Fabric scientist John Skelton said *"Cloth is a mechanism."* With this, he means that the cloth behavior isn't effectively described by a continuum model, but rather by the mechanical interactions of the cloth yarns. As the cloth moves, the yarns collide, bend, and slip causing friction. Inspired by this argument, the discrete model uses a mechanical system of connected particles to model the macroscopic dynamics. Another particle-based method for dynamical simulation was proposed by Provot [1995]. In their work, they use point masses connected by springs to model the elastic behavior of cloth.

Later in the 1990s, Baraff and Witkin [1998] introduced triangle-based cloth simulations using implicit integration. This technique enabled fast simulations of relatively complex clothing. This method is still the foundation for many state-of-the-art implementations used today.

We recommend the excellent books by Volino and Magnenat-Thalmann [2000] and House and Breen [2000] for further reading and a historical background. Another excellent tutorial can be found in the work of Thalmann et al. [2004].

1.5   OVERVIEW OF THIS BOOK

This document explains different approaches to cloth simulation in computer graphics—hopefully, in an understandable way. We will start with a simple approach and work our way up to more complex methods. We will highlight a few different commonly used methods for cloth simulation in production. Focusing on both realistic cloth simulation as well as more approximate but fast methods, better fit for real-time applications. Additionally, we have added a chapter on controlling cloth simulations and future reading.

In order to build a cloth simulator we will first have to make sure that we can store geometry and the cloth simulation state. This is discussed in Chapter 2. Once we can represent the digital garments, we will need to find a way to make them move over time to get dynamic simulations. There are many different mathematical approaches to this and they all have their own advantages and disadvantages. In Chapter 3, we explain the most simple approach which updates the cloth particles in a straightforward way using only information that is available at the current time step.

Once we know how to represent the cloth and how to compute motion, we'll discuss how we can describe a material model for cloth in Chapter 4. This will allow us to compute internal forces that model the cloth behavior. The discussion of handling external collisions is postponed until Chapter 9. A more advanced method to model internal forces more truthfully is described in Chapter 7. Due to the shortcomings of the simple method described in Chapter 3, we will discuss two alternative methods for time integration in Chapters 5 and 6.

As a bonus chapter, we discuss a technique for controlling simulations to obtain art directed results in Chapter 8. This is an advanced topic and is not essential for creating a functional cloth simulator. We conclude the book with references for further reading in Chapter 10.

Beyond the Basics

Some sections will be labeled *Beyond the Basics*. This indicates that understanding this section is not essential for creating a working cloth simulator. These sections provide more advanced information.

1.6   INTENDED AUDIENCE

This book is for anyone interested in learning more about cloth simulation for computer graphics. That being said, we are assuming some background knowledge in solving differential equations and numerical integration. Basic computer graphics knowledge on geometry representation is also assumed to be familiar to the reader. We provide an appendix with commonly used mathematical expressions to help the reader.

1.7   GETTING STARTED

We believe that this book and the references therein contain all the information needed to successfully implement your own cloth simulator. Before you start, we recommend that you first try to set up and run a simulation with an existing simulator. This will give you an introduction to the different steps in the cloth simulation pipeline and the expected behavior of a typical simulation. We recommend the freely available cloth simulator in *Blender*.[1] There are many tutorials available on the web to get you started. Alternatively, for more research oriented readers we recommend the *ARCSim*[2] or *Bullet*[3] cloth simulation libraries.

[1]https://www.blender.org/
[2]http://graphics.berkeley.edu/resources/ARCSim/index.html
[3]https://pybullet.org/wordpress/

Once you've implemented your own solver, we recommend creating a simple square patch of cloth to check for correctness. This simple scene will allow you to test different material properties and constraints. To really validate your implementation we recommend the *Berkeley Garment Library*.[4] This library contains a number of different virtual garments fit for simulation.

[4]http://graphics.berkeley.edu/resources/GarmentLibrary/index.html

CHAPTER 2

Cloth Representation

"All beginnings are difficult."

While this popular saying might be true, we'll try to present you with an understandable explanation.

So you want to learn more about physically based simulation of cloth? That's great! Let's start with the basics. We are assuming you know a little bit about computer graphics already so this chapter will be a very quick overview.

2.1 TRIANGLES

One way to represent geometry on a computer is by using triangles. A single triangle is pretty boring but by combining many triangles into a triangle mesh we have the capability to create astonishingly complex geometrical shapes. Just think about all the special effects you see in movies and video games these days. Almost indistinguishable from real life, except, in real life, things don't blow up as easily as in the movies.

An example of a virtual garment is shown in Figure 2.1. The dress is made up of numerous small triangles. The right figure shows the wireframe of these triangles. The garment has a natural drape over the body of the virtual character thanks to physically based cloth simulation. The garment reacts to external forces such as gravity or wind and moves with the character due to collisions with the body.

A triangle is made up of three *vertices* or *particles*, connected by edges. These terms can be used interchangeably in most settings. Have a look at Figure 2.2 to see what this looks like, particles are shown in red and the triangle is shown in grey. Having a gorgeous 3D model is pretty neat, but you know what's even neater. Having it move! That's what physics-based simulation is all about.

Cloth is a continuous material but in what follows we will work with a discrete particle representation, this will become more clear later in the tutorial. For now, just blindly trust us.

Figure 2.1: The dress has a natural drape on the body of the character thanks to physically based cloth simulation. The garment is made up of multiple triangles which are shown using a black wireframe overlay in the right image. The cloth model and textures were obtained from user *mnphmnmn* on turbosquid: https://www.turbosquid.com/Search/Artists/mnphmnmn.

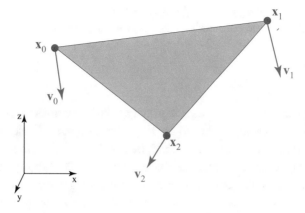

Figure 2.2: A triangle is made up of three particles or vertices with positions $\mathbf{x}_i$ shown in red and velocities $\mathbf{v}_i$ shown in blue.

2.2   PARTICLES

A particle i is defined by a 3D position $\mathbf{x}_i \in \mathbb{R}^3$ and velocity $\mathbf{v}_i \in \mathbb{R}^3$. The combination of particle position and velocity is also referred to as the *particle state* $\mathbf{q}_i = \langle \mathbf{x}_i, \mathbf{v}_i \rangle$. The positions and velocities of the particles will change over time because they have to obey the physical laws that describe the material properties. For cloth, this means that it won't stretch too much but it might shear and bend, creating folds pretty easily. As you know, a wool sweater behaves differently from a linen shirt. This is described by the material model.

We can group all the positions and velocities of the entire particle system with N particles with positions and velocities $\mathbf{x}_i, \mathbf{v}_i \in \mathbb{R}^3$ in a single long vector $\mathbf{x} \in \mathbb{R}^{3N}$ and $\mathbf{v} \in \mathbb{R}^{3N}$

$$
\mathbf{x} = \begin{bmatrix} x_{0_x} \\ x_{0_y} \\ x_{0_z} \\ \vdots \\ x_{N-1_x} \\ x_{N-1_y} \\ x_{N-1_z} \end{bmatrix}, \quad
\mathbf{v} = \begin{bmatrix} v_{0_x} \\ v_{0_y} \\ v_{0_z} \\ \vdots \\ v_{N-1_x} \\ v_{N-1_y} \\ v_{N-1_z} \end{bmatrix}. \tag{2.1}
$$

We can store the triangle meshes using a few different data structures. The most straightforward way is to store the particle positions and velocities in separate arrays that can be indexed using the unique and unchanging particle index i. Triangles are then represented using a list of three integer particle indices per triangle. This is a very simple approach and more complex techniques that encode triangle connectivity can be found in Botsch et al. [2010].

2.3   FORCES

The cloth is affected by external forces such as gravity, wind, or collisions with a body. But, what really makes the cloth behave like it should are the internal forces. These are the stretch, shear, and bend forces that act on the particles to make it behave like textile. Throughout this document we will look at different ways to formulate these internal forces acting on the vertices of the triangles. Once we have these forces we can use numerical integration to advance the simulation over time. That brings us to another important point. We'll be working on a computer and that means we don't have infinite resources available. A typical way to compute these simulations is to only compute the particle states at discrete time steps. Starting from some time t_0, the simulation will advance with small steps of duration h to continue to time $t_0 + h$, $t_0 + 2h$, and onward!

2.3.1   FRAMES AND STEPS

In computer graphics and video, in general the illusion of continuous motion is created by showing the viewer many images per second. The number of images per second is called *frames per second* or *frame rate*. Commonly used frame rates are 24, 30, or even 60 frames per second. This is the number of images we have to create to obtain 1 second of video. However, as we'll see later in this book. We might have to take multiple simulation steps per frame to obtain stable results. The number of simulation steps can thus be much higher than the frame rate. We have to compute the motion of the cloth for every step but we only have to create an image to show the viewer for steps that coincide with frames.

To summarize, we have two types of discretizations in our computer model for cloth:

- **Discretization in space.** The continuous cloth is represented by a finite number of triangles that are made up by particles with positions and velocities.

- **Discretization in time.** Continuous time will be divided into discrete time steps of duration h.

CHAPTER 3

Explicit Integration

"To be wrong as fast as you can is to sign up for aggressive, rapid learning."

Ed Catmull

3.1 INTRODUCTION

Now that we know how to represent clothing using triangles and particles, let's have a look at how we can compute their motion over time. We will start with the most simple approach and then work our way up to more complex algorithms as we go. Time integration is a mathematical technique that is applied to compute how dynamic systems evolve over time.

The state of a particle i is determined by its position $\mathbf{x}_i = \left[x_{i_x}, x_{i_y}, x_{i_z}\right] \in \mathbb{R}^3$ with units [m] and velocity $\mathbf{v}_i = \left[v_{i_x}, v_{i_y}, v_{i_z}\right] \in \mathbb{R}^3$ with units [m/s]. Of course, $\mathbf{x}_i(t)$ and $\mathbf{v}_i(t)$ change over time. Otherwise, our simulations will be pretty boring.

3.2 EXPLICIT INTEGRATION

We know from the laws of motion that velocity is the time derivative of the positions and that acceleration $\mathbf{a}_i = \left[a_{i_x}, a_{i_y}, a_{i_z}\right] \in \mathbb{R}^3$ with units $\left[\frac{\text{m}}{\text{s}^2}\right]$ is the time derivative of the velocities. Let's use the notation for all particles in one vector like in Equation (2.1). Mathematically, this is expressed as

$$\frac{d\mathbf{x}(t)}{dt} = \mathbf{v}(t)$$

$$\frac{d\mathbf{v}(t)}{dt} = \mathbf{a}(t).$$

(3.1)

The most simple way to discretize these equations in time is to use the following finite difference approximation:

$$\frac{\mathbf{x}(t+h) - \mathbf{x}(t)}{h} \approx \mathbf{v}(t)$$

$$\frac{\mathbf{v}(t+h) - \mathbf{v}(t)}{h} \approx \mathbf{a}(t)$$

(3.2)

or after some rewriting

$$\mathbf{x}(t+h) \approx \mathbf{x}(t) + h\mathbf{v}(t)$$
$$\mathbf{v}(t+h) \approx \mathbf{v}(t) + h\mathbf{a}(t),$$

(3.3)

where h is the time step in seconds [s]. From Newton's laws of motion we know that forces $\mathbf{f} \in \mathbb{R}^{3N}$ act as accelerations on the system

$$\mathbf{f}(\mathbf{x}, \mathbf{v}, t) = \mathbf{M}\mathbf{a}(t),$$

(3.4)

looking at the units we see that the unit of force is $\left[\frac{\text{kg·m}}{\text{s}^2}\right]$. This unit is also known as a Newton.

The matrix $\mathbf{M}$ is the mass matrix which is more thoroughly discussed in Chapter 4. In essence, it groups all the individual particle masses m_i into one big matrix representing the entire system.

Equation (3.4) tells us that we have to find all internal and external forces acting on our cloth and use these to accelerate the particle velocities. Forces can be grouped into two categories.

- **Internal forces** are those resulting from the cloth model. These forces will respond to the internal deformations of the cloth such as stretching, compression, shearing, and bending. These deformations will be discussed thoroughly in the next chapter but have a look at Figure 4.1 to get a more concrete understanding.

- **External forces** are, for example, gravity, collisions, or aerodynamic effects such as wind. See the chapter by Ling [2000]. These forces do not result from the cloth material itself. A discussion of collision response is postponed until Chapter 10.

Using discrete time notation and by combining Equations (3.3) and (3.4), we find

$$\begin{aligned}
\mathbf{x}_{n+1} &= \mathbf{x}_n + h\mathbf{v}_n \\
\mathbf{v}_{n+1} &= \mathbf{v}_n + h\mathbf{M}^{-1}\mathbf{f}_n,
\end{aligned} \tag{3.5}$$

where $n + 1$ represents the next time step and n the current one. The matrix $\mathbf{M}^{-1}$ is the inverse of the mass matrix. Practically, this means that for all particles in our cloth simulation, we will have to find all the internal and external forces acting on them. These forces will result in accelerations for the particles. Once we find these accelerations, we can use them to find the new velocities for the next time step. The positions are updated by moving them forward according to their velocity.

In the next chapter, we will have a look at a cloth model that tells us how to compute the internal forces.

What does Equation (3.5) mean exactly? We are advancing the system from one time step to the next, by assuming that forces and velocities are constant during the time step and equal to the force at the beginning of the time step. Of course, in real life this is probably not the case, since it is a continuous time system. However, this is the trade-off we make when we apply a discretization to the system. This will also be the root of the instabilities that might arise when using this integration technique when the time step is too large.

Take a look at Figure 3.1 for a visualization of the time derivative of the position. Is this the best discretization we can come up with? Definitely not. But, it's the simplest one so we will work with this one for now and we'll see more complex discretizations later.

The discretization approach we just derived is known as **Forward Euler Integration** or also **Explicit Integration**. The discretization can be interpreted as the line tangent to the function at the current time and by taking a small step in this direction to advance to the next time step. You can already see from the figure that this is an approximation. Our new state at time $t_1 = t_0 + h$ indicated by the red dot is no longer on the $\mathbf{x}(t)$ curve. The idea is that, if the time step h is not too big, this won't be too much of a problem, … fingers crossed, knock on wood.

3.3   STABILITY ANALYSIS

Let's take some time to discuss the stability properties of the explicit Euler integrator a little bit more. We stated that the explicit Euler method will only be stable for a relatively small time step size h. It would be interesting if we could quantize this more formally.

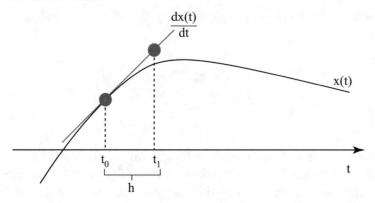

Figure 3.1: Finite difference approximation of the velocity for one time step advancing t_0 to t_1.

3.3.1 TEST EQUATION

We can investigate this restriction by analyzing the following simple initial value problem:

$$\frac{d\ y(t)}{dt} = f(t, y(t))$$

$$y(0) = y_0.$$

(3.6)

In practice, stability analysis of an integrator is not performed on the equations of an actual systems, but rather, on a highly simplified *test equation*. Integrators that perform poorly on this test equation can be discarded without further analysis. It stands to reason that if it doesn't perform well on this simple task, it will probably be even worse for more complex systems. The integrators that do perform well can then be subjected to further analysis. The test equation that we will work with is given by

$$\frac{d\ y(t)}{dt} = \lambda y(t),$$

(3.7)

where $\lambda \in \mathbb{C}$ can be a complex number that is independent of time, $\lambda = \alpha + i\beta$. This test equation is also known as *Dahlquists equation*. In this formulation, i is the imaginary unit which is defined by the property $i^2 = -1$. The analytical solution to this equation is given by

$$y(t) = y_0 e^{\lambda t}.$$

(3.8)

This solution is easily checked, just plug Equation (3.8) into Equation (3.7), given the initial value $y(0) = y_0$. The magnitude of the solution to this differential equation is given by

$$|y(t)| = |y_0| \cdot \left| e^{(\alpha + i\beta)t} \right|. \tag{3.9}$$

The solution of this differential equation is only stable when the real part of λ is non-positive, we denote this by Re (λ). That is the only way the solution will decay over time.

When Re(λ) is positive, the magnitude of the solution will increase with every time step. Most phenomena in the real world don't inject energy and grow to infinity over time so it is okay to assume that the systems we will be trying to solve have bounded behavior as well. All passive physical systems will come to rest. Therefore, from now on, we are assuming that λ will satisfy this requirement of being non-positive.

3.3.2 EXPLICIT EULER ANALYSIS

So far in our stability analysis, we've only talked about our initial value problem and its analytic solution. We haven't talked about any particular time discretization. We will now look at how explicit Euler will approximate this function. The integration is given by

$$y_{k+1} = y_k + hf(t_k, y_k)$$

$$= y_k + h\lambda y_k \tag{3.10}$$

$$= (1 + h\lambda)\, y_k,$$

and by induction, we find the solution at discrete time $t_k = hk$ based on the initial condition y_0 as

$$y_k = (1 + h\lambda)^k \, y_0. \tag{3.11}$$

Given our previous analysis of analytical solution of the continuous time equation, it is not unreasonable to ask that this discretized solution will be bounded as well when time k goes to infinity. The discretized explicit Euler solution will only be bounded when

$$|1 + h\lambda| < 1. \tag{3.12}$$

This is a very interesting find. We see that the behavior of the discrete time solution depends on the positive time step size h. The set of values $h\lambda$ for which the condition holds is called the stability region. Mathematically, this is formulated as

$$S = \{h\lambda \in \mathbb{C} : |1 + h\lambda| < 1\}. \tag{3.13}$$

The stability region is visualized in Figure 3.2. The value λ is determined by the equation itself so that only leaves us with h. Typically, h has to be chosen to be small for the product $h\lambda$ to be in the stability region. This restriction on the time step can be quit severe for our simulations, requiring us to use a very small time step in order to stably advance the simulation. Cloth simulations typically result in stiff equations with $\text{Re}(\lambda)$ negative and great in magnitude.

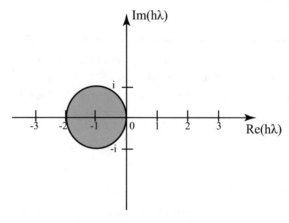

Figure 3.2: Stability region for the explicit Euler method. The horizontal axis is the real axis and the vertical one is the imaginary axis. The method will only produce bounded results when $h\lambda$ is within the stability region shown in blue. The stability region is a unit disk centered at $(-1, 0)$ in the complex plane.

Beyond the Basics

Think about what would happen when we apply explicit Euler integration to an equation where the solutions are concentric circles. Such an equation is given by $y(t) = -\omega^2 \dfrac{d^2 y(t)}{dt}$. The true solution is supposed to orbit forever on the circle it started on. However, due to the discretization, explicit Euler cannot accurately capture this and the solution will always spiral outward no matter how small you take the time step size h. This motivates the need for damping forces. These damping forces are thoroughly explained in Chapter 4.

3.4   ADAPTIVE TIME STEPPING

Beyond the Basics

We performed a short stability analysis and showed that there is quite a restrictive condition imposed on the step size. In any practical application, this restriction will force us to slowly advance the simulation with many small steps. The exact restriction will depend on the cloth material and stiffness constants, this will become clear in the next chapter.

When pursuing this explicit integration approach, it might be interesting to consider adaptive time stepping. Once the simulated results become unstable, the obtained animations will become useless and a new simulation with a smaller time step has to be initiated. In order to avoid having to restart the simulation, we can monitor the error while we perform the integration and vary the step size in order to reduce the risk of instabilities when needed.

Most numerical algorithms for adaptive time stepping focus on accuracy. Instead, in computer graphics, we're mostly interested in stability and only require visual plausibility. In the next chapter, we will see that cloth resists stretching much more than it resists bending or shearing. The stabilities are thus likely to arise due to the effect of the stretch forces. Because of this, Baraff and Witkin [1998] proposed a simple but effective method that monitors the amount of stretching and adapts the step size accordingly.

The simulation is advanced using the current time step size with potential position updates $\Delta \mathbf{x}$. Given these predicted new positions, the resulting stretch forces in the cloth model is computed. When the amount of stretch is beyond a certain limit, then the proposed update is discarded and the step size is halved. New potential positions are then computed using this smaller time step. This limit can be chosen loosely because instabilities will quickly lead to very large position updates.

When the simulation is not on the onset of instability, we would like to increase the time step again so that we can obtain better computational efficiency. When the simulator is able to successfully take multiple steps using a certain step size, the step size is doubled as long as this doesn't make it surpass a user-set maximum

3.5   CONCLUSION

We explained how the continuous time differential equations that define the cloth movement has to be discretized in time in order to be solved using a computer program. The method that was applied to advance the simulation from one time step to the next is called explicit integration.

This is the simplest approach to time integrating the problem. We gather all forces acting on the particles at the current time step and use this to accelerate the particles to obtain a new velocity.

A stability analysis showed that the time step size will have to be chosen small in order to obtain stable results. An adaptive time stepping approach was described to alleviate the restriction. Despite the step size drawback of explicit integration, the method is still attractive from an implementation point of view since we only require information from the current time step to compute the particle states for the next time step. This makes it straightforward to advance the simulation in time using closed-form expressions. A single time step update can be computed rather efficiently. In the next chapter, we will have a look at how the cloth model is defined and how forces are computed.

Unfortunately, in practice, explicit integration is rarely an acceptable choice for advancing the cloth simulation in time and we will have to resort to more complex integration methods such as those described in Chapters 5 and 6.

CHAPTER 4

Mass-Spring Models

"May the force be with you"

Obi-Wan Kenobi; … and a lot of Star Wars fans.

In this chapter, we'll look at how the internal cloth forces are computed.

4.1 INTRODUCTION

You've probably heard somewhere before that all material is constructed out of atoms. As such, it's no surprise that also cloth is made out of atoms and neighboring atoms exert forces on each other preventing excessive stretching or compression. In computer graphics, we can take this idea and represent the continuous cloth by a discrete set of points. Of course, not quite as many as the number of atoms… Unless you're a very patient person! Continuing on this idea of having point masses exerting forces on each other to retain certain properties naturally leads to the mass-spring model.

4.2 COMPUTING MASSES

The name of the model is probably a little bit of a give-away but mass-spring models are none other than point masses connected by springs. Let's say you have some geometry that you would like to simulate as cloth—you can simply take the N vertices of the triangles as the point masses in our simulation model. Besides being a point, point masses have mass.

A good method to determine the mass is to have a surface density ρ with units $\left[\frac{kg}{m^2}\right]$ defined for a material. We model the cloth using 2D triangle elements without thickness. Heavier material will have a higher density and vice versa. We can loop over all the triangles and compute the mass as the triangle surface times the density to obtain the mass of that triangle. This mass is then equally distributed by adding one third of the triangle mass to all three vertices of the triangles. This is an approximation but works well in practice. A single particle will have mass contributions from all triangles it is part of. This area is assumed to be the area in the reference and thus undeformed configuration. This is the configuration the geometry would be in when it is undeformed by forces acting on it.

A commonly used trick is to add some additional mass to the particles that are on the hem of the garments. This extra mass is not necessarily proportional to the triangle area. The reason this results in increased realism is that the hem is often folded double and stitched. We can model this heavier double layer of cloth by modifying the masses. This is much easier than actually representing the geometric fold-over of the hem.

Later in this book, we'll see that in order to express our equations of motion conveniently, we will conceptually have a $\mathbb{R}^{3N \times 3N}$ dimensional mass matrix $\mathbf{M}$. This matrix has the particle masses on the diagonal and is zero otherwise.

Of course, since this matrix has such a simple structure, we only have to store an array of length N with one scalar mass value per particle. The full mass matrix $\mathbf{M}$ will appear in some of the equations that follow. Just to be completely clear, you don't want to construct this as a full dense matrix in your code but it is defined as follows:

$$\mathbf{M} = \begin{bmatrix} m_0 & 0 & 0 & 0 & \cdots & 0 \\ 0 & m_0 & 0 & 0 & \cdots & 0 \\ 0 & 0 & m_0 & 0 & \cdots & 0 \\ 0 & 0 & 0 & m_1 & \cdots & 0 \\ \vdots & \vdots & \vdots & \vdots & \ddots & \vdots \\ 0 & 0 & 0 & 0 & \cdots & m_{N-1} \end{bmatrix}. \tag{4.1}$$

4.3   COMPUTING FORCES

Assuming you're human: from experience with wearing clothes and interacting with textiles in everyday life, we know that cloth is not supposed to stretch or shear all that much. On the other hand, cloth tends to bend out of plane easily creating wrinkles and folds. These types of deformations are visualized in Figure 4.1.

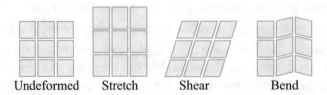

Undeformed   Stretch   Shear   Bend

Figure 4.1: A simple visualization of stretching, shearing, and bending deformations of a square cloth patch.

In order to model resistance to deformations, we can simply construct a spring connecting every pair of neighboring particles. A simple mass-spring system containing nine particles is shown in Figure 4.2. The springs that we just constructed can be thought of as two different types that serve a different purpose. The springs that are shown in green resist stretching of the lattice and the purple springs counteract shearing forces.

A lot of the interesting visual information such as wrinkles and folds results from the cloth bending. A way to incorporate this in our model is to connect 2-ring neighbor particles with a spring, skipping the particle in between. These are called bend springs and are shown in yellow.

It might be interesting to keep these types separated because the spring constant depends on the type of spring. This will make it easier to set material parameters on the model so we have dials to control stretch and shear resistance separately. Most materials have a lower resistance to shearing. Varying the shear stiffness affects the visual behavior dramatically. As a guideline, stretch springs will have very stiff constants whereas shear and bend springs will have small values. Obviously, there is not a complete separation between stretching, shearing, and bending. For example, shear springs will also have some effect on stretching.

Keep in mind that it is totally up to you which particles you connect with springs. Just know that this will eventually have a profound effect on the way the cloth behaves. This will be more clear later on.

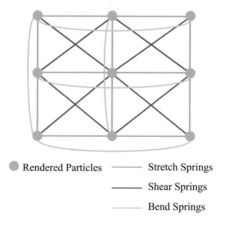

Figure 4.2: A simple mass-spring system consisting of nine particles connected by stretch springs and shear springs. Bend springs connect with every other particle.

4.3.1   ENERGY MINIMIZATION

Physical systems are always trying to reach a minimal energy state. Just think of a marble rolling down a hill, decreasing its potential energy. We can model this is by defining energies and by minimizing them. The forces are those that will try to bring the system in a state that has lower energy until we reach an equilibrium state. This is also known as the second law of thermody-namics.

Now, if you remember from your calculus class, the gradient of a function points in the direction of steepest ascent. But, we don't want to reach a higher energy state, so intuitively, it makes sense for the forces to be the negative gradient of the energy potential function $E(\mathbf{x}) \in \mathbb{R}$

$$\mathbf{f}(\mathbf{x}) = -\frac{\partial E(\mathbf{x})}{\partial \mathbf{x}}. \tag{4.2}$$

Note that this is only the case for *conservative forces*. A force is conservative when the total work done in moving the particle between two points is independent of the path taken. Another way of saying this is that when the particle moves in a loop but starts and ends in the same position, then the net work done will be zero.

These conservative internal forces only depend on the current particle positions. Other forces such as friction and collisions don't. As such, external forces such as collisions or friction forces are not defined by potential energies and are discussed in Chapter 10. To make this more clear, let's start with the most simple example of a conservative force: gravity.

Let's say that in our coordinate system, gravity acts along the z-axis. The gravitational acceleration g is roughly equal to 9.81 $\left[\frac{m}{s^2}\right]$ depending on where in the world you are. The potential energy for a point mass i due to gravity is $E_g(\mathbf{x}_i) = m_i g x_{iz}$ with x_{iz} the component of the particle position along the z-axis. Following Equation (4.2), the resulting force will be

$$\mathbf{f}_i(\mathbf{x}) = -\frac{\partial E_g(\mathbf{x})}{\partial \mathbf{x}_i}$$

$$= -\left[\frac{\partial E_g}{\partial x_{ix}}, \frac{\partial E_g}{\partial x_{iy}}, \frac{\partial E_g}{\partial x_{iz}}\right] \tag{4.3}$$

$$= -\begin{bmatrix} 0 \\ 0 \\ m_i g \end{bmatrix}.$$

That looks a lot like Newton's second law of motion, $\mathbf{f} = \mathbf{M}\mathbf{a}$, doesn't it?

4.3.2 SPRING POTENTIAL ENERGY AND FORCE

Continuing with our cloth simulation, energy is stored in the springs whenever the spring is not at its rest length, i.e., when it is compressed or stretched. *Hooke's law* gives us an expression for determining the potential energy stored in a spring with rest length L and spring constant k with units $[\text{kg/s}^2]$. This value k is also known as the *spring stiffness constant*. It plays an important role as it expresses how much the spring will resist deformation and it provides us with a dial to model different materials.

For a spring connecting particle i and j, we have the potential energy

$$E_{ij}(\mathbf{x}) = \frac{1}{2}k \left(||\mathbf{x}_i - \mathbf{x}_j|| - L\right)^2 .$$ (4.4)

$|| \cdot ||$ is the Euclidian distance; see Equation (A.4). Now that we have our energy function for the spring, we can compute the forces that the spring exerts on particle i and j by taking the derivatives. The force on particle i is found as

$$\mathbf{f}_i(\mathbf{x}) = -\frac{\partial E_{ij}(\mathbf{x})}{\partial \mathbf{x}_i}$$

$$= -k \left(||\mathbf{x}_i - \mathbf{x}_j|| - L\right) \frac{\left(\mathbf{x}_i - \mathbf{x}_j\right)}{||\mathbf{x}_i - \mathbf{x}_j||}$$ (4.5)

and, similarly, the force on particle j is computed as

$$\mathbf{f}_j(\mathbf{x}) = -\frac{\partial E_{ij}(\mathbf{x})}{\partial \mathbf{x}_j}$$

$$= k \left(||\mathbf{x}_i - \mathbf{x}_j|| - L\right) \frac{\left(\mathbf{x}_i - \mathbf{x}_j\right)}{||\mathbf{x}_i - \mathbf{x}_j||}.$$ (4.6)

Looking at Figure 4.3, it should come as no surprise that $\mathbf{f}_i = -\mathbf{f}_j$. A spring connecting two particles will either pull or push on the particles on opposite directions with the same amount of force along the same axes. When the springs are in their rest position, they they will not exert any forces on the point masses. Note that this is a conservative force, just like gravity. The force the spring exerts is independent of the path taken and only depends on the endpoints.

The energy function is quadratic in the particle positions. The force is computed as the negative gradient and will therefore be linear in the positions. The resulting force will scale linearly with the amount of stretching or compression. This is called a linear spring or also, a *Hookean* spring. A graph showing the spring response for different spring stiffnesses k is shown in Figure 4.4.

Figure 4.3: A single spring connecting particle i and j applies equal and opposite forces to the particles along the direction connecting the particles.

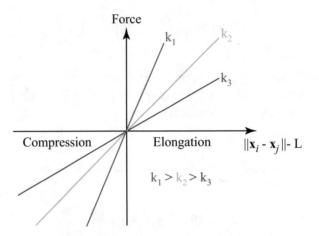

Figure 4.4: The spring force will be linear in the amount of stretching or compression. A larger spring constant will result in a bigger force response for a certain elongation or compression. The horizontal axis shows the deviation from the rest length.

4.3.3   SPRING DAMPING FORCE

Obtaining stable simulation results is critically dependent on having damping forces in the system. We hinted at this when discussing the stability of the oscillatory equation and integration using explicit integration in Section 3.3.2 of the previous chapter. The most simple way to model a damping force for a particle is to add a force that opposes the motion. For a particle i connected to particle j we have the damping force acting on particle i as

$$\mathbf{d}_i(\mathbf{x}) = -k_d\left(\mathbf{v}_i - \mathbf{v}_j\right)$$

$$= -\mathbf{d}_j(\mathbf{x})$$

(4.7)

with k_d the damping coefficient. This mimics the real-world behavior of energy dissipation. Note that this damping model is easy but far from perfect. It prevents bending of the cloth and it penalizes rigid rotations of the spring. Adding a small amount of damping will result in stable simulations. Adding too much damping will make the cloth seem to behave as if it were underwater.

4.4   PUTTING IT ALL TOGETHER

Let's say we start at time t_n. At each step we want to advance our time by a step h. In order to do so we need to compute our update in positions Δx and velocities Δv. Following the discretized Newton's law of motion given in Equation (3.5), we come up with the following system:

$$\Delta x = h v_n$$
$$\Delta v = h \left(M^{-1} f(x_n, v_n) \right) \tag{4.8}$$

with $\Delta x = x_{n+1} - x_n$ and $\Delta v = v_{n+1} - v_n$. For every particle in the system, we can compute all the internal and external forces that are acting upon it and accumulate this in a single force vector. The internal forces are computed as the negative energy gradient and the external forces are added to these internal forces. This will allow us to find the velocity update Δv. Given Δx and Δv, the next state can trivially be found as

$$x_{n+1} = x_n + \Delta x$$
$$v_{n+1} = v_n + \Delta v. \tag{4.9}$$

Phew! Now that we finished all of that, we finally have a working cloth simulator, congratulations! Now is a good time to pat yourself on the back. You might notice however that the results aren't always as great as you hoped. Particularly, the solution might *explode* (not in an awesome special-effects-kind-of-way). The true solution will deviate dramatically from the computed solution unless you take very small time steps. The approximation visualized in Figure 3.1 can be pretty crude. Small discretization errors accumulate and the approximation quickly becomes worse and worse.

4.5   TEARABLE CLOTH

Beyond the Basics

At the beginning of our exposition, we talked about how springs model the internal elastic forces of the cloth. The more the springs are extended, the stronger the resulting force will be. Can we keep stretching the cloth indefinitely? Probably not, right?

Typically, cloth will stretch a small amount without too much resistance. However, stretching beyond this point will result in very strong forces that will resist this deformation. For example, cloth usually doesn't stretch much under its own weight. This can be modeled using advanced techniques which are briefly mentioned in the discussion of this chapter. A different way to handle this is to implement tearable cloth.

Some materials rip when stretched too far. As it turns out, this is actually very easy to model in our simulations and results in interesting dynamic motion. When the springs are stretched a certain fraction too far from their rest length, we can assume the cloth breaks and tears. In our model, we can simply remove this spring from the mass-spring network. This will disconnect the particles in question, creating a tear. This is the most simple approach to the tearing phenomenon. A more advanced method can be found in the work of Metaaphanon et al. [2009].

4.6   OTHER MASS-SPRING APPLICATIONS

Beyond the Basics

In this chapter, we explained how mass-spring systems can be used to model the dynamics of cloth. We wanted to quickly inform you of the fact that mass-spring systems have multiple additional applications in physics-based animation. The focus of this book is cloth simulation so we won't go into too much detail here but we will point you to further reading.

4.6.1   HAIR SIMULATION

Mass-spring systems have been successfully used for the simulation of hair dynamics. The model presented by Selle et al. [2008] incorporates collisions, friction, and torsion and is capable of producing clumping and sticking behavior. Mass-spring systems have also been used by Iben et al. [2013] to generate highly art directed curly hair. The method has proven to be incredibly successful in production.

A simple mass-spring system for hair is shown in Figure 4.5. Note that in addition to the geometric particles there are ghost particles that are necessary to model the hair dynamics. These ghost particle won't be used to render the hair geometry, hence the name.

4.6.2   SOFT BODY DYNAMICS

An extension to three dimensional deformable objects can easily be made. Just like we modeled cloth using triangles where the particles are connected by springs along the edge, we can model deformable volumes using tetrahedra. The geometry is discretized using tetrahedra, representing the full volume. This is also known as a tetrahedralization. A single tetrahedron is visualized in

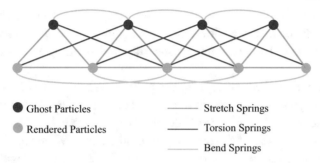

Ghost Particles   Stretch Springs

Rendered Particles   Torsion Springs

    Bend Springs

Figure 4.5: Visualization of a possible mass-spring discretization for a single hair strand. Just like cloth simulations, the dynamics are modeled using a variety of spring connections between particles.

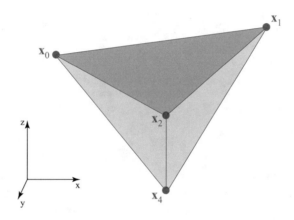

Figure 4.6: Visualization of a single tetrahedron element. The 3D element consist of four connected particles.

Figure 4.6. The element consists of four vertices and the particles are connected using springs along the edges. This will make the tetrahedron want to preserve volume when subjected to external forces. For a more elaborate discussion, we refer to the work of Teschner et al. [2004].

4.7 CONCLUSION

We have introduced the most simple implementation for a cloth solver. All the particle states for the next time step can be computed based on information of the current time step. We presented a method that models the cloth dynamics by using different types of linear springs to incorporate stretching, shearing and bending in the cloth material. The linear spring will have a force response linearly related to the amount of stretch or compression. This makes simulation simple but doesn't realistically reproduce physical cloth behavior. Typically, cloth will be able to stretch a small amount after which it will resist stretching much stronger. This can be modeled using nonlinear springs or piecewise linear springs. More advanced techniques such as strain limiting can also be used. The resistance to shearing and bending is much smaller than the resistance to stretching.

The forces exerted by these springs on the point masses can be computed by thinking of the problem as an energy minimization of the entire particle system. We saw that the forces are thus computed as the negative gradients of the Hookean energies.

Mass-spring models are not the only way to model cloth. More sophisticated techniques using finite element methods are not uncommon in the literature. We provide a more advanced continuum method in Chapter 7.

CHAPTER 5

Implicit Integration

> "I think what children need is love, stability, consistency, and kindness."
>
> *Rosie O'Donnell*

Much like children, cloth simulations need stability and love.

5.1 INTRODUCTION

Artists love big time steps h because it advances the simulation more quickly. Large time steps will decrease accuracy but that might not be a big problem, as long as the results are visually pleasing, we're happy. However, we also saw that when the time step is too large, the simulation results will be unstable and completely unusable.

To obtain more stable simulation results, Baraff and Witkin [1998] had the excellent idea of applying an implicit integration scheme to cloth simulations. This complicates the integration a little bit since we will have to deal with second derivatives. All things considered, the effort is well worth it. The idea is that implicit integration would enable us to obtain stable results, even when using large time steps. This means that even though the computation of one step will be more computationally intensive compared to explicit Euler, we can take larger steps to advance time in the simulation more rapidly without having to worry about stability issues.

5.2 BACKWARD EULER

Let's say we start at time t_n. At each step, we want to compute our update in positions $\Delta x = x_{n+1} - x_n$ and velocities $\Delta v = v_{n+1} - v_n$ using the following integration scheme:

$$\begin{aligned}
\Delta x &= h\left(v_n + \Delta v\right) \\
\Delta v &= h\left(M^{-1}f(x_n + \Delta x, v_n + \Delta v)\right).
\end{aligned} \qquad (5.1)$$

Notice that we are evaluating the forces at the end of the time step. This is the difference with the explicit method we talked about earlier. The time integration given in Equation (5.1) is known as **Implicit Integration** or **Backward Euler**.

In the explicit method, we were evaluating at the current time step where all quantities are known. This allowed us to form a closed expression. In the explicit method, we were just blindly advancing the system with whatever acceleration the forces provided us with at any time step. This is exactly what lead to the stability issues. In what follows, we will see that the linearized backward Euler scheme will at least provide us with a future state which has gradients pointing back to the prior state resulting in much more stable animations; see Figure 5.1. The notation in Equation (5.1) is equivalent to

$$\mathbf{x}_{n+1} = \mathbf{x}_n + h\mathbf{v}_{n+1}$$
$$\mathbf{v}_{n+1} = \mathbf{v}_n + h\mathbf{M}^{-1}\mathbf{f}_{n+1}, \qquad (5.2)$$

where we used the simplified notation $\mathbf{f}_{n+1} = \mathbf{f}(\mathbf{x}_{n+1}, \mathbf{v}_{n+1}) = \mathbf{f}(\mathbf{x}_n + \Delta\mathbf{x}, \mathbf{v}_n + \Delta\mathbf{v})$.

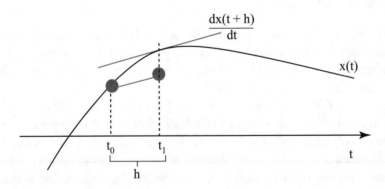

Figure 5.1: Visualization of backward Euler for computing the next particle state.

5.2.1   LINEARIZATION

The forces in Equation (5.2) might be nonlinear and we can accurately solve this using the Newton-Raphson method. However, we can approximate the nonlinear system with a faster (but less accurate) linear system. This is equivalent to taking only one Newton step. Of course, linearizing the equations only once, will result in an approximate solution to the backward Euler formulation. However, it does provide us with a formulation that gives us a close enough answer given the computational cost. In computer graphics, we strive for visual plausibility more than we strive for accuracy so this is a reasonable trade-off.

Linearization is achieved by replacing the nonlinear force term by its first-order Taylor approximation

$$\mathbf{f}(\mathbf{x}_n + \Delta\mathbf{x}, \mathbf{v}_n + \Delta\mathbf{v}) \approx \mathbf{f}_n + \frac{\partial\mathbf{f}}{\partial\mathbf{x}}\Delta\mathbf{x} + \frac{\partial\mathbf{f}}{\partial\mathbf{v}}\Delta\mathbf{v}, \tag{5.3}$$

where $\mathbf{f}_n = \mathbf{f}(\mathbf{x}_n, \mathbf{v}_n)$. Remember, $\mathbf{f}$ and $\mathbf{x}$ are vectors in $\mathbb{R}^{3N}$. This means that $\frac{\partial\mathbf{f}}{\partial\mathbf{x}}$ and $\frac{\partial\mathbf{f}}{\partial\mathbf{v}}$ will be of dimension $\mathbb{R}^{3N \times 3N}$. Substituting Equation (5.3) in the nonlinear Equation (5.1) above, will result in the linear system and by substituting $\Delta\mathbf{x} = h\,(\mathbf{v}_n + \Delta\mathbf{v})$ in the bottom row to eliminate $\Delta\mathbf{x}$, we find the following system for the velocity update

$$\Delta\mathbf{v} = h\mathbf{M}^{-1}\left(\mathbf{f}_n + \frac{\partial\mathbf{f}}{\partial\mathbf{x}}h(\mathbf{v}_n + \Delta\mathbf{v}) + \frac{\partial\mathbf{f}}{\partial\mathbf{v}}\Delta\mathbf{v}\right). \tag{5.4}$$

This is called an implicit method because we see the unknown velocity updates $\Delta\mathbf{v}$ appear on both the left-hand and right-hand side. We can't simply solve for this value by bringing it over to one side of the equation. In order to compute this, we will have to solve a linear system. Reordering the equation above, we obtain the following:

$$\left(\mathbf{I} - h\mathbf{M}^{-1}\frac{\partial\mathbf{f}}{\partial\mathbf{v}} - h^2\mathbf{M}^{-1}\frac{\partial\mathbf{f}}{\partial\mathbf{x}}\right)\Delta\mathbf{v} = h\mathbf{M}^{-1}\left(\mathbf{f}_n + h\frac{\partial\mathbf{f}}{\partial\mathbf{x}}\mathbf{v}_n\right), \tag{5.5}$$

with $\mathbf{I} \in \mathbb{R}^{3N \times 3N}$ being the identity matrix. When constructing these matrices we will compute $\frac{\partial\mathbf{f}}{\partial\mathbf{x}}$ and $\frac{\partial\mathbf{f}}{\partial\mathbf{v}}$ for the internal forces but all external forces are just grouped in $\mathbf{f}_n$ and don't necessarily have derivatives contributing to the system matrix unless we know how to compute them.

Equation (5.5) is a linear system of the form $\mathbf{A}\Delta\mathbf{v} = \mathbf{b}$. If you remember from your linear algebra class, there's many ways to solve this for $\Delta\mathbf{v}$. Later in this chapter we will discuss one approach but know that there are options.

Now, this matrix $\mathbf{A}$ isn't just any old matrix. This matrix is actually a sparse matrix that has a certain block structure, we will go into more detail later. Keep in mind for now that, although technically you could store everything as a dense matrix and solve a dense system, this is most definitely not the most efficient implementation. We give a more suitable matrix representation in Section 5.5. An efficient way to solve this system is explained in Section 5.7.

5.3   STABILITY ANALYSIS

We can investigate the stability properties of this implicit Euler integration by looking at the same test equation, as discussed in Chapter 3. Analog to before, we discretize the continuous time equation. This time, let's look at the behavior of the backward Euler scheme. The discretization of the test equation is given by

$$y_{k+1} = y_k + hf(t_{k+1}, y_{k+1})$$
$$= y_k + h\lambda y_{k+1} \tag{5.6}$$

or after grouping terms, we find

$$(1 - h\lambda)\, y_{k+1} = y_k. \tag{5.7}$$

The next time step is then computed as

$$y_{k+1} = \frac{1}{(1 - h\lambda)} y_k. \tag{5.8}$$

Just like before, induction brings us to the following expression:

$$y_k = \left(\frac{1}{(1 - h\lambda)}\right)^k y_0. \tag{5.9}$$

We assumed that the exact solution of the equations we are solving for will be bounded when time goes to infinity. This was expressed using the condition that $\text{Re}(h\lambda)$ is non-positive. The time step h is always positive so this is equivalent to $\text{Re}(\lambda)$ being non-positive. The requirement for the discretized solution to be bounded is

$$\left|\frac{1}{1 - h\lambda}\right| < 1, \quad \text{Re}(\lambda) < 0. \tag{5.10}$$

The remarkable thing is that this condition is satisfied for any positive time step h and real λ. The implicit Euler method will be unconditionally stable. The stability of the results of explicit Euler depends heavily on the chosen time step size. This is clearly not the case for implicit integration. It is worth noting that obtaining stable results is not the same as obtaining accurate results. It merely means that the discretized solution of a stable differential equation will remain bounded when simulation time goes to infinity.

A visualization of the stability region is shown in Figure 5.2. The set defining the stability region is given by

$$S = \{h\lambda \in \mathbb{C} : |1 - h\lambda| > 1\}. \tag{5.11}$$

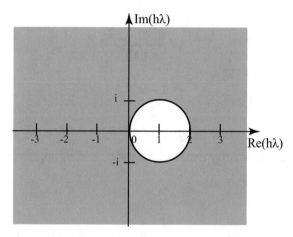

Figure 5.2: The stability region in the complex plane for the implicit Euler method is highlighted in blue. The horizontal axis is the real axis and the vertical one is the imaginary axis. For $h\lambda$ in the negative complex plane corresponds to stable systems for which the implicit Euler method is unconditionally stable. The positive half plane (except for the unit circle at $(1, 0)$) corresponds to unbounded systems for which backward Euler will obtain bounded solutions. This is sometimes referred to as *overstability*. For values lying in the circle in the right plane, the system is unbounded and so will be the numerical solution.

5.4 SPRING FORCES AND THEIR DERIVATIVES

For an explicit solver, we only needed to compute and apply the forces associated with the springs and the external forces. We just saw that for an implicit solver, we will also need the internal force derivatives with respect to the positions and velocities.

For sake of clarity, we will use the shorthand notation $\mathbf{x}_{ij} = \mathbf{x}_i - \mathbf{x}_j$. A normalized vector is denoted by a hat $\hat{\mathbf{x}}$. Recall that the force $\mathbf{f}_s$ exerted on particle i resulting from a spring connecting particle i and j was computed as

$$\mathbf{f}_s = -k\left(||\mathbf{x}_{ij}|| - L\right)\hat{\mathbf{x}}_{ij}. \tag{5.12}$$

Let's take the derivative with respect to the first particle i. Recall that the following equality holds:

$$\frac{\partial\hat{\mathbf{x}}}{\partial\mathbf{x}} = \frac{\partial\left(\frac{\mathbf{x}}{||\mathbf{x}||}\right)}{\partial\mathbf{x}}$$

$$= \frac{\mathbf{I}||\mathbf{x}|| - \mathbf{x}\hat{\mathbf{x}}^T}{||\mathbf{x}||^2} \tag{5.13}$$

$$= \frac{\mathbf{I} - \hat{\mathbf{x}}\hat{\mathbf{x}}^T}{||\mathbf{x}||}$$

with $\mathbf{I}$ the 3×3 dimensional identity matrix. Using this, and by applying the chain rule we find $\frac{\partial\mathbf{f}_s}{\partial\mathbf{x}} \in \mathbb{R}^{3\times3}$

$$\frac{\partial\mathbf{f}_s}{\partial\mathbf{x}_i} = -k\left(\left(||\mathbf{x}_{ij}|| - L\right)\frac{\partial\hat{\mathbf{x}}_{ij}}{\partial\mathbf{x}_i} + \hat{\mathbf{x}}_{ij}\frac{\partial\left(||\mathbf{x}_{ij}|| - L\right)}{\partial\mathbf{x}_i}\right)$$

$$= -k\left(\left(||\mathbf{x}_{ij}|| - L\right)\left(\frac{\mathbf{I} - \hat{\mathbf{x}}_{ij}\hat{\mathbf{x}}_{ij}^T}{||\mathbf{x}_{ij}||}\right) + \hat{\mathbf{x}}_{ij}\hat{\mathbf{x}}_{ij}^T\right) \tag{5.14}$$

$$= -k\left(\left(1 - \frac{L}{||\mathbf{x}_{ij}||}\right)\left(\mathbf{I} - \hat{\mathbf{x}}_{ij}\hat{\mathbf{x}}_{ij}^T\right) + \hat{\mathbf{x}}_{ij}\hat{\mathbf{x}}_{ij}^T\right).$$

Looking at Figure 4.3, unsurprisingly, we find

$$\frac{\partial\mathbf{f}_s}{\partial\mathbf{x}_i} = -\frac{\partial\mathbf{f}_s}{\partial\mathbf{x}_j}. \tag{5.15}$$

These 3×3 dimensional blocks will only be non-zero when particle i and j are connected. For every spring, there will be four blocks in the global $\dfrac{\partial \mathbf{f}}{\partial \mathbf{x}}$ matrix. There's the force acting on particle i with respect to the derivative of $\mathbf{x}_i$ and $\mathbf{x}_j$, and also the force acting on particle j with respect to the derivative of $\mathbf{x}_i$ and $\mathbf{x}_j$. Summarized, for a spring connecting particle i and j, we can intuitively state that:

- index(i,i): Expresses how the force on particle i changes when the $\mathbf{x}_i$ changes;
- index(i,j): Expresses how the force on particle i changes when the $\mathbf{x}_j$ changes;
- index(j,i): Expresses how the force on particle j changes when the $\mathbf{x}_i$ changes; and
- index(j,j): Expresses how the force on particle j changes when the $\mathbf{x}_j$ changes.

Conceptually, the full Jacobian matrix $\dfrac{\partial \mathbf{f}}{\partial \mathbf{x}} \in \mathbb{R}^{3N \times 3N}$ will be a sparse symmetric matrix that might look like this

$$
\frac{\partial \mathbf{f}}{\partial \mathbf{x}} =
\begin{bmatrix}
\square & \square & \mathbf{0} & \mathbf{0} & \cdots & \square \\
\square & \square & \mathbf{0} & \square & \cdots & \mathbf{0} \\
\mathbf{0} & \mathbf{0} & \square & \mathbf{0} & \cdots & \mathbf{0} \\
\mathbf{0} & \square & \mathbf{0} & \square & \cdots & \square \\
\vdots & \vdots & \vdots & \vdots & \ddots & \vdots \\
\square & \mathbf{0} & \mathbf{0} & \square & \cdots & \square
\end{bmatrix},
\tag{5.16}
$$

where each $\square$ represents a 3×3 block and $\mathbf{0}$ represents a 3×3 block of zeros. Remember that $\mathbf{f} = -\dfrac{\partial E}{\partial \mathbf{x}}$, so we find that the Jacobian matrix is symmetric since

$$
\frac{\partial^2 E}{\partial \mathbf{x}_i \, \partial \mathbf{x}_j} =
\begin{bmatrix}
\dfrac{\partial^2 E}{\partial x_{ix} \partial x_{jx}} & \dfrac{\partial^2 E}{\partial x_{ix} \partial x_{jy}} & \dfrac{\partial^2 E}{\partial x_{ix} \partial x_{jz}} \\[2.5ex]
\dfrac{\partial^2 E}{\partial x_{iy} \partial x_{jx}} & \dfrac{\partial^2 E}{\partial x_{iy} \partial x_{jy}} & \dfrac{\partial^2 E}{\partial x_{iy} \partial x_{jz}} \\[2.5ex]
\dfrac{\partial^2 E}{\partial x_{iz} \partial x_{jx}} & \dfrac{\partial^2 E}{\partial x_{iz} \partial x_{jy}} & \dfrac{\partial^2 E}{\partial x_{iz} \partial x_{jz}}
\end{bmatrix}
= \left(\frac{\partial^2 E}{\partial \mathbf{x}_j \, \partial \mathbf{x}_i} \right)^T.
\tag{5.17}
$$

Row i of this matrix is computed by looking at particle i and all other particles j connected to i. For every connected particle pair, there will be a contribution at location (i, i), (j, j), (j, i), and (i, j). The diagonal elements of the full matrix will be a sum of multiple contributions of all the different springs connected to that particle. The off-diagonal elements will only have a single contribution assuming that there's only one spring connecting that specific pair of particles.

We can easily find that the Jacobian $\frac{\partial \mathbf{d}_i}{\partial \mathbf{v}_j} \in \mathbb{R}^{3\times3}$ of the damping force is

$$\frac{\partial \mathbf{d}_i}{\partial \mathbf{v}_j} = -k_d \mathbf{I}. \tag{5.18}$$

5.5   BLOCK COMPRESSED ROW STORAGE

Beyond the Basics

You might have noticed that these matrices can get very big for high resolution cloth geometry. Detailed cloth meshes will result in very large matrices. A lot of the entries will be zero though and it doesn't make much sense to store all these zeros in a dense matrix. This type of matrix that has only a few non-zero elements is called a sparse matrix and we should represent it as such on the computer for efficiency reasons.

There are many ways of doing this and there's a vast amount of literature out there published by researchers who have spent a lot of time figuring out the best way to do this. Here, we will focus on an approach called *Block Compressed Row Storage* (BCRS). We will store all the non-zero blocks per row of the system matrix.

Since we're working in three dimensions all positions and velocities are represented as 3D vectors. For N particles this means that all positions and velocities can be stored in a $\mathbb{R}^{3N}$ vector for each. In our code we'll loop over all particles frequently so it is convenient to have an N dimensional array where every element is a small $\mathbb{R}^3$ vector representing position or velocities or any other quantity associated with a particle. The matrix in Equation (5.5) is of dimension $3N \times 3N$. However, it will be more convenient to index between 0 and $N-1$ so that every element (i, j) represents a 3×3 dimensional block. Here is a simple didactic visualization of the type of matrix we'll be working with

$$\begin{bmatrix} \boxed{0} & \boxed{1} & 0 & 0 & \boxed{2} \\ \boxed{1}^T & 0 & \boxed{3} & 0 & 0 \\ 0 & \boxed{3}^T & 0 & \boxed{4} & \boxed{5} \\ 0 & 0 & \boxed{4}^T & 0 & \boxed{6} \\ \boxed{2}^T & 0 & \boxed{5}^T & \boxed{6}^T & 0 \end{bmatrix}. \tag{5.19}$$

Note that for cloth simulations there will be contributions on all diagonal elements of the matrix. Since the matrices will be square and symmetric, we will actually only have to store the upper triangle half of the matrix. All the other data can be deferred from this. We just have to remember that there's a transposed block on the other side of the diagonal when we're multiplying with the matrix. This is only a performance issue and not a correctness issue, so we will continue our explanation with the full sparse matrix. Just keep this in mind when you're implementing your own data structure.

It's easy to get distracted reading long boring mathematical texts, but now is a good time to pay attention. The essence of the BCRS format is described in this paragraph. The BCRS format works by storing the blocks sequentially per row in an array that we name blocks. Feel free to be creative in naming your data structures. Additionally, we need to keep a list of the associated column indices so we can figure out where in the full matrix the blocks belong. We call this array columnIndex. The only missing information now is how many blocks there are in each row. We encode this in an array named rowPointer by storing the accumulative number of blocks per row. This means that row i has rowPointer $[i + 1] -$ rowPointer $[i]$ number of blocks.

The easiest way to understand, is to look at an example. The BCRS format for the matrix in Equation (5.19) is stored using the following arrays:

$$\begin{array}{rl} \text{blocks}: & \left[\ \boxed{0}\ \boxed{1}\ \boxed{2}\ \boxed{1}\ \boxed{3}\ \boxed{3}\ \boxed{4}\ \boxed{5}\ \boxed{4}\ \boxed{6}\ \boxed{2}\ \boxed{5}\ \boxed{6}\ \right] \\ \text{columnIndex}: & \left[\ 0\ \ 1\ \ 4\ \ 0\ \ 2\ \ 1\ \ 3\ \ 4\ \ 2\ \ 4\ \ 0\ \ 2\ \ 3\ \right] \\ \text{rowPointer}: & \left[\ 0\ \ 3\ \ 5\ \ 8\ \ 10\ \ 13\ \right]. \end{array}$$

Note that we're only storing non-zero blocks in our data array named blocks. Not wasting any memory here! Well ..., actually ..., except for the fact that, for clarity of explanation, we're not exploiting the symmetry of course.

5.5.1 MATRIX-VECTOR MULTIPLICATION

We now have an efficient way of storing our huge but sparse matrices in memory. The next question that arises is: how do we perform a matrix-vector multiplication using this data structure? All shall be explained.

We can process the rows of the BCRS matrix in parallel for a multiplication $\mathbf{b} = \mathbf{Xa}$ since we know that element $\mathbf{b}[i]$ of the output is the multiplication of row i with the input vector $\mathbf{a}$ of the input vector.

For every row, we loop over the blocks in that row and add the result of the multiplication of the block with the input element to the output. Pseudocode of the algorithm is given in Algorithm 5.1.

Algorithm 5.1 BCRS Matrix-Vector Multiplication

1: **for** $i = 1$ **to** rowPointer.size() $- 1$ **do**
2:    const int row = i-1;
3:    const int blocksInThisRow = rowPointer[i] - rowPointer[row];
4:    **for** $j = 0$ **to** blocksInThisRow $- 1$ **do**
5:       const int blockIndex = rowPointer[row] + j;
6:       const int column = columnIndex[blockIndex];
7:       output[row] += blocks[blockIndex]*rhs[column];
8:    **end for**
9: **end for**
10: **return** output

5.6 ADDING VELOCITY CONSTRAINTS

We are getting close to having a fully implicit solver for mass-spring systems. One thing that we should add is the ability to constrain particles.

Imagine you want to simulate clothes fluttering on a clothes line. We need a way to attach the clothes to the clothes line and to make sure it stays attached during the simulation. This is done by computing a filter matrix $\mathbf{S}_i \in \mathbb{R}^{3\times 3}$ for every particle that is constrained. This matrix will restrict the movement either fully, i.e., the particle isn't allowed to move at all. Alternatively, the particle can be constrained to move in a plane or along a specified axis. The number of degrees of freedom of a particle i is denoted as ndof(i).

The particle can be prohibited to move in a unit direction $\mathbf{p}$ by applying the filter operation where ndof(i)=2 in Equation (5.20). Similarly, the particle is prevented to move in two orthonormal directions $\mathbf{p}$ and $\mathbf{q}$ by applying the matrix defined for ndof(i)=1 in Equation (5.20).

We'll tell you how to build this matrices here. For them to have effect on the simulation they will have to be applied during the modified conjugate gradient solver which is described in the next subsection.

Without further ado, the filter matrix will be constructed as

$$
\mathbf{S}_i = \begin{cases} \mathbf{I} & \text{if ndof}(i) = 3 \\[2mm] (\mathbf{I} - \mathbf{p}_i \mathbf{p}_i^T) & \text{if ndof}(i) = 2 \\[2mm] (\mathbf{I} - \mathbf{p}_i \mathbf{p}_i^T - \mathbf{q}_i \mathbf{q}_i^T) & \text{if ndof}(i) = 1 \\[2mm] \mathbf{0} & \text{if ndof}(i) = 0 \end{cases} .
\tag{5.20}
$$

The first and last case are pretty trivial. Multiplying with the identity matrix won't do anything and multiplying by all zeros will eliminate all movement for the particle. It should also be clear that we only need to construct and store filter matrices for the constrained particles. It would be inefficient to store a separate matrix for every unconstrained particle since this multiplication is the identity operation.

The above filter matrices will constrain the particles to have zero accelerations in the specified directions. In addition to this there is also a way to exactly specify the change in velocity for a particle. This is achieved by introducing a new vector $\mathbf{z}_i$ for every particle i. In the next section, we will see how we can solve the linear system to obtain velocity updates that correspond to the filter operations and the new constraint variable $\Delta \mathbf{v}_i = \mathbf{z}_i$.

5.7   SOLVING THE LINEAR SYSTEM

The linear system will be solved using a modified preconditioned conjugate gradient solver. The method is modified because we will perform the filtering operations to incorporate the constraints during the conjugate gradient solve as proposed by Baraff and Witkin [1998]. We refer to the paper by Shewchuk [1994] for a very excellent introduction to the conjugate gradient method.

Remember, the system we are solving for the velocity updates was given by Equation (5.5). Note that the left-hand side matrix will only be symmetric when all the particles have equal mass. We can make the system symmetric, regardless of the particle masses, by left multiplying with the mass matrix $\mathbf{M}$. We find the following system:

$$
\left(\mathbf{M} - h \frac{\partial \mathbf{f}}{\partial \mathbf{v}} - h^2 \frac{\partial \mathbf{f}}{\partial \mathbf{x}} \right) \Delta \mathbf{v} = h \left(\mathbf{f}_n + h \frac{\partial \mathbf{f}}{\partial \mathbf{x}} \mathbf{v}_n \right).
\tag{5.21}
$$

This formulation allows us to efficiently use a preconditioned conjugate gradient method. This method is particularly well suited for positive definite symmetric systems. The left-hand side matrix $\mathbf{A}$ of the system is stored in the BCRS format. The right-hand side $\mathbf{b}$ is a dense vector. They are defined as follows:

$$\mathbf{A} = \left(\mathbf{M} - h\frac{\partial \mathbf{f}}{\partial \mathbf{v}} - h^2 \frac{\partial \mathbf{f}}{\partial \mathbf{x}} \right)$$

$$\mathbf{b} = h\left(\mathbf{f}_n + h\frac{\partial \mathbf{v}}{\partial \mathbf{x}} \mathbf{v}_n \right).$$

(5.22)

The conjugate gradient algorithm used to iteratively solve this system $\mathbf{A}\Delta\mathbf{v} = \mathbf{b}$ is given in Algorithm 5.2. It is an update method that starts with an initial guess that is iteratively updated by adding scalar multiples of the search directions. The filtering algorithm to implement the velocity constraints is given in Algorithm 5.3.

After converging, the solution will satisfy the following two conditions.

- For each particle, the component of the residual vector $\mathbf{r}$ in the unconstrained directions will be zero.

- For each particle, the component of $\Delta\mathbf{v}$ in the constrained directions will equal the prescribed constraint $\mathbf{z}$.

5.7.1   PRECONDITIONING

Preconditioning is a technique that is commonly used to transform the system to a form that's more suitable for a numerical algorithm. This is the reason Krylov methods have such good properties. In our case we would like the system matrix to be close to the unity matrix since would make the system trivial to solve. Decreasing the condition number of the matrix will increase the rate of convergence. The preconditioning matrix $\mathbf{P}$ that we use is a simple diagonal matrix that is readily available and inexpensive to compute. The diagonal elements are computed as $\mathbf{P}_{ii} = \dfrac{1}{\mathbf{A}_{ii}}$, this is also known as diagonal scaling.

Alternative preconditioners are for example incomplete Cholesky factorization, successive-symmetric over-relaxation or block diagonal preconditioners.

5.8   POSITION ALTERATIONS

So far, we've seen how constraints can be used to impose conditions on the particle positions. It seems natural to also want to impose constraints on the particle positions. A common example is when a cloth particle collides with a solid object and needs to be displaced to be back on the

---

Algorithm 5.2 Modified Preconditioned Conjugate Gradients

1: $\Delta v = z$
2: $\delta_0 = \mathbf{filter(b)}^T \, \mathbf{P} \, \mathbf{filter(b)}$
3: $\mathbf{r} = \mathbf{filter(b - A\Delta v)}$
4: $\mathbf{c} = \mathbf{filter(P^{-1}r)}$
5: $\delta_{new} = \mathbf{r}^T \mathbf{c}$
6: **while** $\delta_{new} > \epsilon^2 \delta_0$ **do**
7:   $\mathbf{q} = \mathbf{filter(Ac)}$
8:   $\alpha = \delta_{new} / \left(\mathbf{c}^T \mathbf{q}\right)$
9:   $\Delta v = \Delta v + \alpha \mathbf{c}$
10:   $\mathbf{r} = \mathbf{r} - \alpha \mathbf{q}$
11:   $\mathbf{s} = \mathbf{P}^{-1}\mathbf{r}$
12:   $\delta_{old} = \delta_{new}$
13:   $\delta_{new} = \mathbf{r}^T \mathbf{s}$
14:   $\mathbf{c} = \mathbf{filter}(\mathbf{s} + \frac{\delta_{new}}{\delta_{old}}\mathbf{c})$
15: **end while**

Algorithm 5.3 Constraint Filter

1: **for** $i = 1$ **to** N **do**
2:   $\hat{\mathbf{a}}_i = \mathbf{S}_i \mathbf{a}_i$
3: **end for**
4: **return** $\hat{\mathbf{a}}$

---

object boundary. You could just displace the particles during the simulation but this would lead to instabilities since the neighboring particles aren't informed until the next time step. Particles are likely to end up in unfavorable positions, resulting in large forces. In order to make position alterations, we will need to incorporate this update in the entire system update. We can do this by modifying the position update, including the desired displacement $\mathbf{y}_n$ at time t_n as follows:

$$\Delta \mathbf{x} = h \left(\mathbf{v}_n + \Delta \mathbf{v}\right) + \mathbf{y}_n. \tag{5.23}$$

Repeating the derivations made earlier in this chapter, we find the following system:

$$\left(\mathbf{M} - h \frac{\partial \mathbf{f}}{\partial \mathbf{v}} - h^2 \frac{\partial \mathbf{f}}{\partial \mathbf{x}}\right) \Delta \mathbf{v} = h \left(\mathbf{f}_n + h \frac{\partial \mathbf{f}}{\partial \mathbf{x}} \mathbf{v}_n + \frac{\partial \mathbf{f}}{\partial \mathbf{x}} \mathbf{y}_n\right). \tag{5.24}$$

5.9   A QUICK NOTE ON STABILITY

Beyond the Basics

Academic papers unfortunately often leave out important implementation details. Implicit integration does provide numerically stable integration, but the unconstrained global system on the left hand side must satisfy certain properties.

David Eberle suggested that I note the following:

"The unconstrained global system on the left-hand side must always be positive definite. This means that the negative of the local force Jacobians must be semidefinite under all deformation modes. Figuring out which terms of the Jacobian could violate this and devising ways to modify them is a necessary exercise to create a production capable implementation. Choi and Ko [2002] discusses the necessary modification for a simple linear spring force."

5.10   ALTERNATIVE INTEGRATION SCHEMES

Beyond the Basics

We discussed both a fully explicit and a fully implicit integration technique. Explicit integration proved to be too unstable for practical use. Implicit integration as used by Baraff and Witkin [1998] enabled large time steps by linearizing the nonlinear system. That way the integration could be solved efficiently using a conjugate gradient method. Alternatively, Desbrun et al. [1999] used the implicit method but they didn't linearize the system. Instead, the authors split the system in a linear and a nonlinear part. They solve the linear part of the equations but they don't integrate the nonlinear part. This nonlinear term is instead accounted for using a correction term.

These are just a few options out of a vast number of techniques. Specifically, a second-order backward difference method results in more accurate solutions with less damping at a negligible additional cost and similar stability compared to backward Euler.

A semi-implicit integration technique with a second order backward difference formula has successfully been used by Eberhardt et al. [2000] and Choi and Ko [2002]. The integration is given by

$$\frac{1}{h}\left(\frac{3}{2}\mathbf{x}_{n+1} - 2\mathbf{x}_n + \frac{1}{2}\mathbf{x}_{n-1}\right) = \mathbf{v}_{n+1}$$

$$\frac{1}{h}\left(\frac{3}{2}\mathbf{v}_{n+1} - 2\mathbf{v}_n + \frac{1}{2}\mathbf{v}_{n-1}\right) = \mathbf{M}^{-1}\mathbf{f}_{n+1}.$$

(5.25)

The nonlinear force term is discretized like we've already seen in Equation (5.3). Rearranging and grouping terms will result in a system that can be solved to find either $\Delta\mathbf{x}$ or $\Delta\mathbf{v}$ depending on the way the equations are combined. For completeness, the system that solves for $\Delta\mathbf{x}$ is given by

$$\left(\mathbf{I} - h\frac{2}{3}\mathbf{M}^{-1}\frac{\partial\mathbf{f}}{\partial\mathbf{v}} - h^2\frac{4}{9}\mathbf{M}^{-1}\frac{\partial\mathbf{f}}{\partial\mathbf{x}}\right)\Delta\mathbf{x} = \frac{1}{3}(\mathbf{x}_n - \mathbf{x}_{n-1}) + \frac{h}{9}(8\mathbf{v}_n - 2\mathbf{v}_{n-1})$$

$$+ \frac{4h^2}{9}\mathbf{M}^{-1}\left(\mathbf{f}_n - \frac{\partial\mathbf{f}}{\partial\mathbf{v}}\mathbf{v}_n\right) - \frac{2h}{9}\mathbf{M}^{-1}\frac{\partial\mathbf{f}}{\partial\mathbf{v}}(\mathbf{x}_n - \mathbf{x}_{n-1})$$

(5.26)

with $\Delta\mathbf{x} = \mathbf{x}_{n+1} - \mathbf{x}_n$. This is again a sparse symmetric matrix that can be solved using a conjugate gradient method. As a final example, a fourth-order Runge-Kutta method has been applied to cloth simulation by Eberhardt et al. [1996].

5.11 CONCLUSION

This ends our discussion of implicit integration for cloth simulations using mass-spring systems. We have shown how forces can be computed as the negative gradient of potential energies. These forces will accelerate the particles. This approach results in much more stable simulations compared to explicit methods which allows us to take larger time steps at the cost of more computation time. Implicit integration will require more work since we will have to solve a linearized system. The system can be stored in block compressed row storage format for an efficient in memory representation. The system was then iteratively solved using a modified conjugate gradient solver that allows us to implement constraints on the particles.

Despite the additional cost, having stable simulations is extremely important in computer graphics. Unstable results are completely unusable. For this reason, you will almost always want to opt for something more complex than a fully explicit solver. Explicit Euler typically overestimates the energy of the true solution resulting in unstable simulations. In contrast, implicit

Euler will typically underestimate the energy of the true solution resulting in an overly damped look of the simulations.

Damping might not seem to be too big of a problem since damping is a phenomenon that occurs naturally in the world. A big problem is that the amount of damping cannot be explicitly controlled and depends on the resolution, time step and stiffness of the system. Higher-order methods will result in better accuracy and less damping but require more computations to advance the simulation. We refer the interested reader to the work of Hauth [2003] and Dinev et al. [2018] for more information.

CHAPTER 6

Simulation as an Optimization Problem

"I want results and I want them yesterday!"

Your client

Sometimes it is more important to have a simulation result ready in time rather than to have highly accurate results.

Let's dive into a different way to solve mass-spring systems and have a look at the excellent work of Liu et al. [2013]. The authors present a very fast way to simulate mass-spring systems while keeping the results very stable. Just like an implicit solver, but at a much better computational efficiency.

6.1 INTRODUCTION

Interactive applications are very common these days and require the virtual world to be updated at high frame rates in order to be perceived as smooth motion. Possible applications are video games, virtual and augmented reality, and virtual surgery. It is very important to honor this time restriction because not doing so will create a lagging motion that could induce motion sickness in the users of virtual reality applications.

Typically this update is required every 33 ms or even less. This means that we only have a small fixed amount of time available to us to come up with an adequate solution. The real world is even more restrictive because we don't just need to compute physics during the total frame time. Other components such as rendering, networking, and human-computer interaction will also consume a significant amount of this available time.

We start by reformulating the simulation as an optimization problem. Although this reformulation isn't necessarily new, Liu et al. [2013], then proceeded to propose some very smart techniques to speed up the optimization.

6.2   NOTATION

Let's start with some notation. Just like before, we have N particles with concatenated position vector $\mathbf{x} \in \mathbb{R}^{3N}$ and velocity vector $\mathbf{v} \in \mathbb{R}^{3N}$. The particles are connected in a mass-spring network. We again assume conservative forces derived from an energy potential, $\mathbf{f} = -\dfrac{\partial E}{\partial \mathbf{x}}$. The mass matrix $\mathbf{M}$ is the same as defined before in Equation (4.1). Now is a good time to introduce a new notation for the mass matrix. We will be able to use this type of notation for other equations in this chapter too. The mass matrix can be rewritten in terms of a *Kronecker* product which is denoted by the symbol $\otimes$:

$$\mathbf{M} = \tilde{m} \otimes \mathbf{I}. \tag{6.1}$$

This simply means that every element in the diagonal matrix

$$\tilde{m} = \mathrm{diag}\,(m_0, m_1, m_2, \ldots, m_{N-1}) \in \mathbb{R}^{N \times N}$$

will be multiplied by the identity matrix $\mathbf{I} \in \mathbb{R}^{3 \times 3}$ resulting in the mass matrix $\mathbf{M} \in \mathbb{R}^{3N \times 3N}$. More formally, for a matrix $\mathbf{A} \in \mathbb{R}^{m \times n}$ and a matrix $\mathbf{B} \in \mathbb{R}^{p \times q}$ we have

$$\mathbf{A} \otimes \mathbf{B} = \begin{bmatrix} A_{0,0}\mathbf{B} & \cdots & A_{0,n-1}\mathbf{B} \\ \vdots & \ddots & \vdots \\ A_{m-1,0}\mathbf{B} & \cdots & A_{m-1,n-1}\mathbf{B} \end{bmatrix} \in \mathbb{R}^{mp \times nq}. \tag{6.2}$$

6.3   REFORMULATING THE PROBLEM

From Chapter 5, we remember that implicit Euler integration is formulated as

$$\begin{aligned} \mathbf{x}_{n+1} &= \mathbf{x}_n + h\mathbf{v}_{n+1} \\ \mathbf{v}_{n+1} &= \mathbf{v}_n + h\mathbf{M}^{-1}\mathbf{f}_{n+1}. \end{aligned} \tag{6.3}$$

We will now start rewriting this in order to come to an expression that can be converted into an optimization problem. It's just a different approach to solving the time integration of the system. In our case for cloth simulation, we will see that there are some significant advantages when you want to obtain fast results. Let's begin by reformulating the problem. It is clear that the following holds:

$$hv_n = x_n - x_{n-1}$$
$$hv_{n+1} = x_{n+1} - x_n.$$

(6.4)

If we multiply the velocity update in Equation (6.3) with the time step h, we find $hv_{n+1} = hv_n + h^2M^{-1}f_{n+1}$. Plugging in the equalities from Equation (6.4), we find

$$x_{n+1} - x_n = x_n - x_{n-1} + h^2M^{-1}f_{n+1}$$
$$\Leftrightarrow x_{n+1} - 2x_n + x_{n-1} = h^2M^{-1}f_{n+1}.$$

(6.5)

The left-hand side is now actually the finite differences expression for the second derivative of x. It should be very clear by now that if we bring the h^2 and mass term to the left-hand side, we have nothing else than Newton's second law of motion again. For clarity, let's group the known terms so it will be easier to see that we are trying to compute the next particle state x_{n+1}. Let's define $y = 2x_n - x_{n-1}$. To keep things simple, we will just write x to mean x_{n+1}. Putting this all together, we have

$$x - y = h^2M^{-1}f(x).$$

(6.6)

It might seem like the grouping of the terms in y is arbitrary. However, we can see some physical interpretation for this term. It will become very clear if we write it as follows:

$$y = 2x_n - x_{n-1} = x_n + (x_n - x_{n-1}) = x_n + hv_n.$$

(6.7)

This is exactly where the positions of all the particles would move in the absence of forces using an explicit Euler step. In essence, this is Newton's first law: if there are no forces acting on the system, then the system will just keep moving with the current velocities. This is visualized in Figure 6.1 and is called inertia. Therefore, y is sometimes referred to as the inertia term. Now, of course, in our system there are forces so we'll have to take these into account.

6.4   SOLVING THE NONLINEAR ACTUATIONS

From the previous subsection we found the following system of nonlinear actuations where we have $\mathbb{R}^{3N}$ vectors on both side of the equation. The nonlinearity comes from the force term.

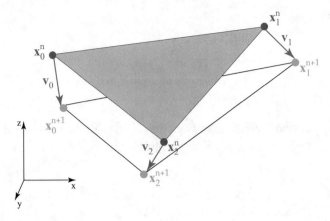

Figure 6.1: Visualization of the inertia update not taking into account the internal and external forces acting on the cloth.

Multiplying Equation (6.6) with the mass matrix $\mathbf{M}$, we find

$$\mathbf{M}(\mathbf{x} - \mathbf{y}) = h^2 \mathbf{f}(\mathbf{x}). \tag{6.8}$$

Now here comes an important realization. Solving Equation (6.8) is equivalent to finding the critical points for which $\dfrac{\partial g(\mathbf{x})}{\partial \mathbf{x}} = 0$ of the following function:

$$g(\mathbf{x}) : \mathbb{R}^{3N} \to \mathbb{R}$$

$$g(\mathbf{x}) = \frac{1}{2}(\mathbf{x} - \mathbf{y})^T \mathbf{M}(\mathbf{x} - \mathbf{y}) + h^2 E(\mathbf{x}). \tag{6.9}$$

This is obvious since setting the gradient of $g(\mathbf{x})$ to zero will give us Equation (6.8) again— Remember that $\dfrac{\partial E(\mathbf{x})}{\partial \mathbf{x}} = -\mathbf{f}(\mathbf{x})$.

We know that critical points correspond to local minima or maxima of a function. Using this information we can reformulate the system in Equation (6.8) as

$$\operatorname*{argmin}_{\mathbf{x}} \; g(\mathbf{x}). \tag{6.10}$$

Therefore, we are looking for the value of $\mathbf{x}$ that minimizes $g(\mathbf{x})$. We do not really care about the actual value of the function. We only care about the minimizer since this solves our original problem of time integrating the mass-spring system.

We are now faced with minimizing a nonlinear function. The straightforward way to go about this is to use Newton's method. If you recall, that is exactly the approach we took in Chapter 5 where we solved implicit integration by using one iteration of Newton's method.

This time, however, we will be taking a different approach, one that will lead to a more efficient implementation when you really care about getting a result on screen fast and aren't too concerned about accuracy.

6.5 LOCAL-GLOBAL ALTERNATION PROBLEM FORMULATION

The method that we will use to optimize this problem formulation is named *local-global alternation*. This method is sometimes also referred to as *block coordinate descent*. The trick to make this strategy work is to introduce some additional auxiliary variables. In our case, these auxiliary variables will be the spring directions. Let's look at Hooke's law again to make this more concrete. For a spring connecting two particles with positions $\mathbf{p}_1$ and $\mathbf{p}_2$ and with rest length L and spring stiffness k, we have the energy definition

$$E(\mathbf{p}_1, \mathbf{p}_2) = \frac{k}{2} \left(||\mathbf{p}_1 - \mathbf{p}_2|| - L \right)^2. \tag{6.11}$$

Making use of the following lemma (see the original paper by Liu et al. [2013] for a proof of this lemma):

Lemma: $\forall \, \mathbf{p}_1, \mathbf{p}_2 \in \mathbb{R}^3$ and $\forall L \geq 0$:

$$\min_{||\mathbf{d}||=L, \mathbf{d}\in\mathbb{R}^3} || (\mathbf{p}_1 - \mathbf{p}_2) - \mathbf{d}||^2 = (||\mathbf{p}_1 - \mathbf{p}_2|| - L)^2. \tag{6.12}$$

The left-hand side is a small minimization problem over the auxiliary variable $\mathbf{d}$, where the positions $\mathbf{p}_1$ and $\mathbf{p}_2$ are kept fixed. The auxiliary variable $\mathbf{d}$ is the vector that represents the spring direction and is constraint to be of length equal to the rest length L. This is easily visualized in Figure 6.2. So, why do we need to be dealing with this additional mathematical construct? It will helps us to rewrite $E(\mathbf{x})$ in Equation (6.9). This way the minimization problem $g(\mathbf{x})$ can be written as a new problem $\tilde{g}(\mathbf{x}, \mathbf{d})$ and additional constraints on the auxiliary variables.

So let's now look at the whole mass-spring system containing S springs. For a spring $i \in [0, \ldots, S-1]$, we have endpoints $\mathbf{p}_{i_1}$ and $\mathbf{p}_{i_2}$ with particle indices i_1 and i_2, respectively.

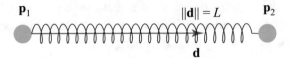

Figure 6.2: The auxiliary variable $\mathbf{d} \in \mathbb{R}^3$ for the spring connecting $\mathbf{p}_1$ and $\mathbf{p}_2$ is shown as the red vector. This variable is equal to the spring direction with length equal to the spring rest length.

Finally, we have a stiffness constant k_i. Using the lemma we can rewrite the energy potential for all springs as

$$\frac{1}{2} \sum_{i=0}^{S-1} k_i \|\mathbf{p}_{i_1} - \mathbf{p}_{i_2} - \mathbf{d}_i\|^2, \tag{6.13}$$

where $\mathbf{d}_i$ is still restricted to have a length equal to the rest length $\|\mathbf{d}_i\| = L_i$.

We can get rid of the sum notation by using a little bit of mathematical trickery. The equation can be rewritten in matrix form

$$\frac{1}{2} \sum_{i=0}^{S-1} k_i \|\mathbf{p}_{i_1} - \mathbf{p}_{i_2} - \mathbf{d}_i\|^2 = \frac{1}{2}\mathbf{x}^T \mathbf{L} \mathbf{x} - \mathbf{x}^T \mathbf{J} \mathbf{d}, \tag{6.14}$$

where $\mathbf{x} \in \mathbb{R}^{3N}$ is the vector that stacks all N particle positions $\mathbf{p}_i \in \mathbb{R}^3$ in a single long vector. Similarly, $\mathbf{d} \in \mathbb{R}^{3S}$ stacks all the individual $\mathbf{d}_i \in \mathbb{R}^3$ vectors. We see that $\mathbf{L} \in \mathbb{R}^{3N \times 3N}$ and $\mathbf{J} \in \mathbb{R}^{3N \times 3S}$ and they are given by the following expressions:

$$\mathbf{L} = \left(\sum_{i=0}^{S-1} k_i \mathbf{A}_i \mathbf{A}_i^T \right) \otimes \mathbf{I}$$

$$\mathbf{J} = \left(\sum_{i=0}^{S-1} k_i \mathbf{A}_i \mathbf{S}_i^T \right) \otimes \mathbf{I}, \tag{6.15}$$

where $\mathbf{I} \in \mathbb{R}^{3 \times 3}$ is the identity matrix. The vectors $\mathbf{A}_i \in \mathbb{R}^N$ and $\mathbf{S}_i \in \mathbb{R}^S$ are indicator functions that are mostly zero. For a spring i, $\mathbf{A}_i$ will have element i_1 equal to 1 and element i_2 equal to

-1. For $\mathbf{S}_i$, element i will be equal to 1:

$$
\mathbf{A}_i = \begin{bmatrix} \vdots \\ 1 \\ \vdots \\ -1 \\ \vdots \end{bmatrix}, \qquad \mathbf{S}_i = \begin{bmatrix} \vdots \\ 1 \\ \vdots \end{bmatrix}. \tag{6.16}
$$

You can write out the matrix equation to convince yourself that these are equivalent. It's also important to realize that these matrices $\mathbf{L}$ and $\mathbf{J}$ are constant in time. As long as the cloth structure doesn't change, the matrices won't change as the particle positions are updated throughout the simulation. This means that we will have to recompute the matrices when we want to model tearing of cloth.

Let's summarize our findings for the energy function so far:

$$
E(\mathbf{x}) = \sum_{i=0}^{S-1} \frac{1}{2} \left(||\mathbf{p}_{i_1} - \mathbf{p}_{i_2}|| - L_i \right)^2
$$

$$
= \min_{\mathbf{d} \in U} \ \frac{1}{2}\mathbf{x}^T \mathbf{L}\mathbf{x} - \mathbf{x}^T \mathbf{J}\mathbf{d} \tag{6.17}
$$

$$
U = \left\{ (\mathbf{d}_0, \dots, \mathbf{d}_{S-1}) \in \mathbb{R}^{3S} : ||\mathbf{d}_i|| = L_i \right\}.
$$

This completes our reformulation of the internal forces due to the springs in the mass spring network. The energy due to the total forces is easily computed by adding a term for the external forces $\mathbf{x}^T \mathbf{f}_{ext}$ to the energy. The derivative with respect to $\mathbf{x}$ of this term will give us just the external forces so our formulation is still equivalent to Equation (6.8).

We finally find that the full reformulation is given by the mimimization over $\mathbf{x}$ and $\mathbf{d} \in U$ of the following function:

$$
\tilde{g}(\mathbf{x}, \mathbf{d}) = \frac{1}{2}(\mathbf{x} - \mathbf{y})^T \mathbf{M}(\mathbf{x} - \mathbf{y}) + \frac{1}{2}h^2 \mathbf{x}^T \mathbf{L}\mathbf{x} - h^2 \mathbf{x}^T \mathbf{J}\mathbf{d} + h^2 \mathbf{x}^T \mathbf{f}_{ext}. \tag{6.18}
$$

6.6   SOLVING TIME INTEGRATION USING LOCAL-GLOBAL ALTERNATION

We have spent quite a bit of time deriving the formulation of $\tilde{g}$ to make sure you understand how it works. If you are only interested in coding up a working implementation, this is where things get interesting.

The local-global optimization consists of two steps that are iteratively executed until convergence or until you decide the solution is good enough. The nice thing is that the error decreases monotonically so you'll know for sure that the more iterations you spend, the more accurate the solution will be. In the first step, we will find the minimizing values for the auxiliary variables **d** assuming that **x** is fixed. This can be done in parallel since every spring can be optimized over seperately. In the second, global step, we will assume **d** is fixed and optimize for **x**.

1. **Local step.** Assume **x** is fixed and optimize for **d**. This can be done in parallel because every spring can be treated separately. This will reset the springs to their rest length. By doing so they break the connection between the particle positions.

2. **Global step.** In the second global step, we assume **d** is fixed and we optimize over all particle states **x** essentially reconnecting the springs into a state with lower energy.

6.6.1   LOCAL STEP

Assuming **x** is fixed, finding the values for **d** is actually very easy. The minimizing values for **d** is just the rest length direction of the spring. We just need to project every spring onto the rest length. This is visualized in Figure 6.3 for a triangle with three springs. Every spring i is rescaled along $\mathbf{p}_{i_1} - \mathbf{p}_{i_2}$ to have length L_i. The figure shows the projected springs in blue. We see that this step disconnects the springs from the particles. We're not changing the directions, only the lengths. This separation will be resolved in the global step which will reconnect the springs to the particles. The projection is performed separately for every spring so this is very easy to parallelize in the implementation. Mathematically, for every spring i we compute $\mathbf{d}_i$ as

$$\mathbf{d}_i = \frac{\mathbf{p}_{i_1} - \mathbf{p}_{i_2}}{||\mathbf{p}_{i_1} - \mathbf{p}_{i_2}||} L_i. \tag{6.19}$$

6.6.2   GLOBAL STEP

In the global step, we will keep **d** fixed and optimize over **x**. Whereas the local step disconnected the springs from the particles, the global step will bring everything together again. In this step, we're left with solving the unconstrained quadratic function given in Equation (6.18) over **x**.

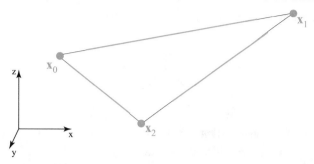

Figure 6.3: Visualization of the local solve where the springs are projected onto their rest length along the direction connecting the two particles. The original springs are shown in black and the projected springs are shown in blue.

We know that the minimum of a quadratic function is simply solved as the critical point. So we find the solution by setting the gradient equal to zero and solving for $\mathbf{x}$

$$\left(\mathbf{M} + h^2\mathbf{L}\right)\mathbf{x} = \mathbf{M}\mathbf{y} + h^2\mathbf{J}\mathbf{d} + h^2\mathbf{f}_{ext}$$

$$\Leftrightarrow \left(\mathbf{M} + h^2\mathbf{L}\right)\mathbf{x} = \mathbf{b}.$$

(6.20)

And this is just a linear system of the form $\mathbf{A}\mathbf{x} = \mathbf{b}$ which we know how to solve. Okay! Phew, we're done, finally! We have now described an alternative way to integrate mass spring systems over time. But wait. We said this was supposed to be a faster method than before, but we still have to solve a linear system in every time step. And now we even have those auxiliary variables to compute! Ahah, very good comment. But the thing that makes this method fast is the fact that both $\mathbf{M}$ and $\mathbf{L}$ are fixed and don't change over time. The right-hand side $\mathbf{b}$ will be different in every time step but the matrix on the left-hand side won't ever change!

For a symmetric positive definite matrix such as our Hessian, a sparse Cholesky factorization is guaranteed to exist and can be precomputed to efficiently solve the linear system in every iteration of the local-global optimization process. The Cholesky factorization of a matrix $\mathbf{A}$ will give you

$$\mathbf{A} = \mathbf{K}\mathbf{K}^T,$$

(6.21)

where $\mathbf{K}$ is a lower triangular matrix meaning that everything above the diagonal will be zero. The system $\mathbf{A}\mathbf{x} = \mathbf{b}$ then becomes $\mathbf{K}\mathbf{K}^T\mathbf{x} = \mathbf{b}$. Defining $\mathbf{K}^T\mathbf{x} = \mathbf{z}$, the overall solution can be found by solving the following two linear systems sequentially

$$\mathbf{K}\mathbf{z} = \mathbf{b}$$
$$\mathbf{K}^T\mathbf{x} = \mathbf{z}.$$

(6.22)

This can be computed very efficiently since $\mathbf{K}$ is a lower triangular matrix that can be precomputed during the initialization of the simulation.

6.7   CONCLUSION

In this chapter, we looked at an alternative approach to solving the time integration for cloth simulation. The integration is reformulated as an optimization problem that can be solved using a two-step approach. The optimization method is called local-global alternation or block coordinate descent. It provides very efficient results because the left-hand side matrix of the linear system can be precomputed and pre-factorized into a very efficient formulation, as long as the particle connectivity doesn't change.

So how does it compare to Newton's method? Using this method, it will be very fast to compute a single iteration of local-global alternation. This allows us to perform many iterations at the same computational cost of a single Newton iteration. This is particularly handy if you have a limited time budget like in video games that demand a specific frame rate.

After a few iterations, there's a cutoff point where Newton's method performs much better than this method. This makes it clear that if you're going for accuracy, Newton's method is the way to go. If, however, you have a limited computation time available to advance the simulation to the next time step, this method might prove to be of great value to you.

CHAPTER 7

Continuum Approach to Cloth

"Give me stability or give me death!"

David Baraff

Baraff's quote is a play on Patrick Henry's historical quote. In this chapter, we will look at a more complex approach to model cloth.

7.1 INTRODUCTION

Modeling garments using a mass-spring system might seem very convenient and straightforward but it has one very big downside: the behavior of cloth actually heavily depends on how you connect the points using springs. Think about it: you decide where the springs are constructed, there's no connection to real-world physics involved.

Let's say you want to recreate your own clothes right now. You would have a very hard time finding good spring connections and stiffness constants that accurately represent certain materials such as linen or nylon. This makes it frustrating, time-consuming, and unintuitive to create nice folds and other distinct properties that make your clothes look the way they do.

The take-home message is that mass-spring systems are conceptually easy and will give you pleasing results **but,** it is very hard to model these virtual garments to match real examples and materials. Breen et al. [1994] pioneered with a method that explicitly represents the microstructure of woven cloth. Later, Baraff and Witkin [1998] introduced a ground-breaking technique that is detailed in this chapter.

7.2 CLOTH REST SHAPE

In essence, mass-spring systems model the cloth by connecting points with lines, but there's no other material in between. You can think of this as almost a 1D representation. It seems only natural to take this a step further and instead look at triangles as a whole instead of just the edges connecting the vertices. Instead of looking at how a spring is compressed or stretched, we will look at how a triangle is stretched and compressed.

Just like springs have a rest length to represent the rest configuration, triangles will need something similar. Since it is a 2D representation we will need two numbers for every particle which are typically named (u, v) that map a vertex into a 2D undeformed rest configuration. This is essentially the shape that the cloth would want to attain if it wasn't subjected to any forces. If you're familiar with computer graphics, this is exactly the same idea as (u, v) mapping for textures. The 3D geometry is unfolded onto a 2D surface on which textures can be painted. Every vertex of the mesh relates to a corresponding point on the (u, v) map.

A very simple example is shown in Figure 7.1. Here, a T-shirt is cut in a front and a back part so it can be unfolded onto the 2D (u, v) space. The blue dotted lines show some of the vertex correspondences and the red lines show where the shirt is sown together.

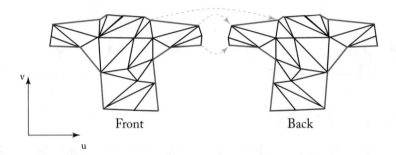

Figure 7.1: Example of a (u, v) map for a simple low resolution T-shirt. The 3D mesh is cut into front and back sections, so it can be unfolded onto the 2D (u, v) space. A few particle correspondences are shown with blue dotted lines and the red lines indicate the edges of triangles that are sewed together in the 3D mesh.

Having (u, v) maps naturally leads to a technique called flat panelling. Flat panelling refers to the tailoring process where flat pieces of cloth are sewn together to create garments. They will always have the tendency to unfold back into this flat panel, but of course they can't ..., because they're stitched. Now, we do have the option to have a rest configuration that doesn't want to be flat by imposing rest-bend angles. This will be clear when discussing bend forces. The (u, v) map for the dress example of Figure 2.1 is shown in Figure 7.2.

7.3 COMPUTING FORCES AND THEIR DERIVATIVES

Just like in Section 4.3, we'll compute forces as the negative gradients of energy functions. For the continuum approach, these energies are based on the triangle as a whole and not just the edges as was the case in the mass-spring network. The energy functions $E_c(\mathbf{x})$ are defined based

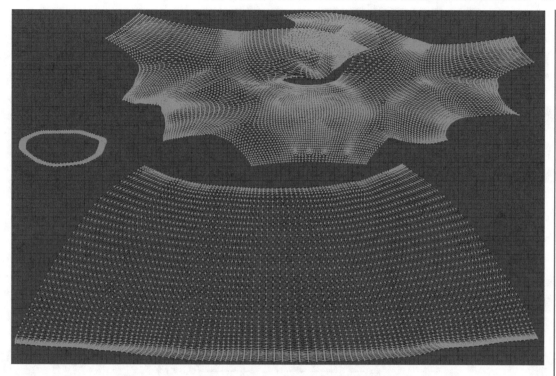

Figure 7.2: Example of a (u, v) map for the dress example shown in Figure 2.1.

on condition functions $\mathbf{C}(\mathbf{x})$ as proposed by Baraff and Witkin [1998]:

$$E_c(\mathbf{x}) = \frac{k}{2}\mathbf{C}(\mathbf{x})^T\mathbf{C}(\mathbf{x}), \tag{7.1}$$

where k is a stiffness constant of our choice. The value of this parameter will determinate the behavior of the modeled material. More specifically, for a condition $\mathbf{C}(\mathbf{x})$ we obtain a force $\mathbf{f}_i \in \mathbb{R}^3$ for particle i:

$$\mathbf{f}_i = -\frac{\partial E_c}{\partial \mathbf{x}_i}$$

$$= -\left[\frac{\partial E_c}{\partial x_{ix}}, \frac{\partial E_c}{\partial x_{iy}}, \frac{\partial E_c}{\partial x_{iz}}\right] \tag{7.2}$$

$$= -k\frac{\partial \mathbf{C}(\mathbf{x})}{\partial \mathbf{x}_i}\mathbf{C}(\mathbf{x}).$$

We have seen earlier that for an implicit solver, we won't just need the forces but also the force derivatives with respect to positions and velocities. Taking a second partial derivative of the above $\mathbf{f}_i$ with respect to particle j gives us a Jacobian block $\mathbf{K}_{ij} \in \mathbb{R}^{3 \times 3}$:

$$\mathbf{K}_{ij} = \frac{\partial \mathbf{f}_i}{\partial \mathbf{x}_j}$$

$$= -k\left(\frac{\partial \mathbf{C}(\mathbf{x})}{\partial \mathbf{x}_i}\frac{\partial \mathbf{C}(\mathbf{x})}{\partial \mathbf{x}_j}^T + \frac{\partial^2 \mathbf{C}(\mathbf{x})}{\partial \mathbf{x}_i \partial \mathbf{x}_j}\mathbf{C}(\mathbf{x})\right)$$

$$= \begin{bmatrix} \dfrac{\partial f_{ix}}{\partial x_{jx}} & \dfrac{\partial f_{ix}}{\partial x_{jy}} & \dfrac{\partial f_{ix}}{\partial x_{jz}} \\[2mm] \dfrac{\partial f_{iy}}{\partial x_{jx}} & \dfrac{\partial f_{iy}}{\partial x_{jy}} & \dfrac{\partial f_{iy}}{\partial x_{jz}} \\[2mm] \dfrac{\partial f_{iz}}{\partial x_{jx}} & \dfrac{\partial f_{iz}}{\partial x_{jy}} & \dfrac{\partial f_{iz}}{\partial x_{jz}} \end{bmatrix}. \tag{7.3}$$

Since K_{ij} is the second derivative of the energy function E_c, we have

$$K_{ij} = \frac{\partial^2 E_c}{\partial \mathbf{x}_i \partial \mathbf{x}_j} = \frac{\partial^2 E_c}{\partial \mathbf{x}_j \partial \mathbf{x}_i} = K_{ji}^T. \tag{7.4}$$

This means that, just like we found in Equation (5.17), the global Jacobian matrix is symmetric.

7.3.1   DAMPING FORCES

To obtain stable and robust cloth simulations, we will also need to apply damping forces to the system. In the real world, energy dissipates in a number of ways and we need to mimic this in our simulator. The damping force $\mathbf{d} \in \mathbb{R}^3$ is defined using the condition function as follows:

$$\mathbf{d} = -k_d \frac{\partial \mathbf{C}(\mathbf{x})}{\partial \mathbf{x}} \dot{C}(x),\qquad(7.5)$$

where k_d is the damping stiffness that we are free to pick. We can thus apply damping forces associated with the conditions imposed on the triangles.

Just like normal forces, these damping forces will also contribute to the force derivatives matrices $\frac{\partial \mathbf{d}_i}{\partial \mathbf{x}_j} \in \mathbb{R}^{3 \times 3}$:

$$\frac{\partial \mathbf{d}_i}{\partial \mathbf{x}_j} = -k_d \left(\frac{\partial \mathbf{C}(\mathbf{x})}{\partial \mathbf{x}_i} \frac{\partial \dot{C}(x)^T}{\partial \mathbf{x}_j} + \frac{\partial^2 \mathbf{C}(\mathbf{x})}{\partial \mathbf{x}_i \partial \mathbf{x}_j} \dot{C}(x) \right).\qquad(7.6)$$

Following the findings of Pritchard [2006], we compute $\dot{\mathbf{C}}(\mathbf{x}) \in \mathbb{R}^3$ as follows:

$$\dot{\mathbf{C}}(\mathbf{x}) = \frac{d\,\mathbf{C}(\mathbf{x})}{dt}$$

$$= \frac{\partial \mathbf{C}(\mathbf{x})}{\partial \mathbf{x}} \frac{\partial \mathbf{x}}{\partial t}\qquad(7.7)$$

$$= \sum_i \left(\frac{\mathbf{C}(\mathbf{x})}{\partial \mathbf{x}_i} \right) \mathbf{v}_i$$

with the sum over all the particles i participating in the condition function. Now pay close attention! Up until now, all matrices were symmetric. We see in Equation (7.6) that the first term breaks symmetry. However, the math dictates that it should be there. This complicates solving the linear system since we saw that it is very advantageous to have symmetric matrices.

One way to overcome this difficulty is to just drop this first term so we can maintain symmetry. Now, we're deviating from the true mathematical model and this is not physically justifiable, but it turns out that the results remain very good.

It's good to keep in mind that in computer graphics, we're not necessarily concerned with getting a super accurate result. Rather, we prefer having a believable and physically plausible image on screen in a reasonable amount of time. Nobody likes waiting, right?

So, going forward, we'll simplify Equation (7.6) by omitting the non-symmetric part. The equation reduces to the following:

$$\frac{\partial \mathbf{d}_i}{\partial \mathbf{x}_j} = -k_d \left(\frac{\partial^2 \mathbf{C}(\mathbf{x})}{\partial \mathbf{x}_i \, \partial \mathbf{x}_j} \left(\frac{\partial \mathbf{C}(\mathbf{x})}{\partial \mathbf{x}_j} \right)^T \mathbf{v}_j \right). \tag{7.8}$$

We still need the derivatives of the damping forces with respect to the velocities $\dfrac{\partial \mathbf{d}_i}{\partial \mathbf{v}_j} \in \mathbb{R}^{3\times 3}$:

$$\frac{\partial \mathbf{d}_i}{\partial \mathbf{v}_j} = -k_d \frac{\partial \mathbf{C}(\mathbf{x})}{\partial \mathbf{x}_i} \frac{\partial \dot{\mathbf{C}}(x)}{\partial \mathbf{v}_j}. \tag{7.9}$$

In the above equation, we will have to compute the derivate $\dfrac{\partial \dot{\mathbf{C}}(x)}{\partial \mathbf{v}_j}$. We know $\dot{\mathbf{C}}(\mathbf{x}) = (\partial \mathbf{C}(\mathbf{x})/\partial \mathbf{x})^T \mathbf{v}$. Taking the derivative with respect to the velocity gives us $\dfrac{\dot{\mathbf{C}}(x)}{\partial \mathbf{v}} \in \mathbb{R}^{3\times 3}$

$$\frac{\dot{\mathbf{C}}(x)}{\partial \mathbf{v}} = \frac{\partial}{\partial \mathbf{v}} \left(\frac{\partial \mathbf{C}(\mathbf{x})^T}{\partial \mathbf{x}} \mathbf{v} \right) \tag{7.10}$$

$$= \frac{\partial \mathbf{C}(\mathbf{x})}{\partial \mathbf{x}}.$$

Putting it all together, by combining Equation (7.9) and (7.10), we find

$$\frac{\partial \mathbf{d}_i}{\partial \mathbf{v}_j} = -k_d \frac{\partial \mathbf{C}(\mathbf{x})}{\partial \mathbf{x}_i} \frac{\partial \mathbf{C}(\mathbf{x})^T}{\partial \mathbf{x}_j}. \tag{7.11}$$

We now know how to compute all the forces and their derivatives given a condition function imposed on the triangles.

In the next sections, we'll take a look at the actual condition functions used for simulating cloth. We will have separate conditions for stretch, shear, and bending.

7.4 STRETCH FORCES

Cloth will get stretched and compressed a little when subjected to forces. A visualization is shown in Figure 7.4. At the beginning of this chapter, we mentioned the existence of a reference configuration of the cloth at rest. This is encoded in the (u, v) map that is specified for the geometry.

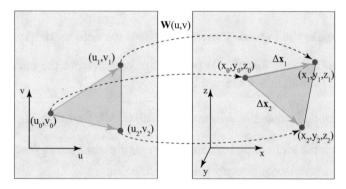

Figure 7.3: A visualization how (u, v) coordinates are mapped to the corresponding 3D positions in the world space using the mapping $\mathbf{W}(u, v)$.

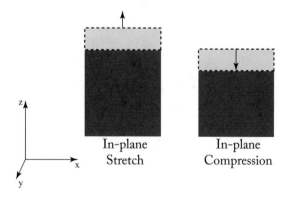

Figure 7.4: Visualization of in plane stretching and compression for two triangles.

The deformation map $\mathbf{W}(u, v, t)$ maps points from the rest configuration (u, v) space to the world space at time t. Let's make this more concrete and look at two neighboring particles $\bar{\mathbf{x}}_1$ and $\bar{\mathbf{x}}_2$ in the (u, v) material space such that $d\bar{\mathbf{x}}_{12} = \bar{\mathbf{x}}_2 - \bar{\mathbf{x}}_1$ is of infinitesimal length. We have

$$\mathbf{x}_1 = \mathbf{W}(\bar{\mathbf{x}}_1), \quad \mathbf{x}_2 = \mathbf{W}(\bar{\mathbf{x}}_1 + d\bar{\mathbf{x}}_{12}) \text{ and } d\mathbf{x}_{12} = \mathbf{x}_2 - \mathbf{x}_1. \tag{7.12}$$

Using a Taylor expansion where we only consider the linear terms we find

$$d\mathbf{x}_{12} = \mathbf{W}(\bar{\mathbf{x}}_1 + d\bar{\mathbf{x}}_{12}) - \mathbf{W}(\bar{\mathbf{x}}_1) \approx \mathbf{W}(\bar{\mathbf{x}}_1) + \frac{\partial \mathbf{W}}{\partial \bar{\mathbf{x}}} d\bar{\mathbf{x}}_{12} - \mathbf{W}(\bar{\mathbf{x}}_1) = \frac{\partial \mathbf{W}}{\partial \bar{\mathbf{x}}} d\bar{\mathbf{x}}_{12}. \tag{7.13}$$

We can thus make the following statements about the deformation map $\mathbf{W}(u, v)$.

- The continuous deformation map $\mathbf{W}(u, v)$ maps **points** from the undeformed 2D (u, v) space to the 3D simulation space, see Figure 7.3.

- **Vectors** are mapped using the *deformation gradient*. We find that $\mathbf{W}_u = \dfrac{\partial \mathbf{W}(u, v)}{\partial u}$ measures how much the u direction is stretched or compressed. Analogously, $\mathbf{W}_v = \dfrac{\partial \mathbf{W}(u, v)}{\partial v}$ measures the stretch or compression in the v direction.

The triangle will be undeformed in the u or the v direction when $||\mathbf{W}_u|| = 1$ or $||\mathbf{W}_v|| = 1$, respectively.

This all sounds very abstract so let's take a look at what this means for a single deformed triangle. We define the following quantities based on the simulation space positions $\mathbf{x}_i$ and rest configurations (u_i, v_i):

$$\begin{aligned}
\Delta\mathbf{x}_1 &= \mathbf{x}_1 - \mathbf{x}_0 \\
\Delta\mathbf{x}_2 &= \mathbf{x}_2 - \mathbf{x}_0 \\
\Delta u_1 &= u_1 - u_0 \\
\Delta u_2 &= u_2 - u_0 \\
\Delta v_1 &= v_1 - v_0 \\
\Delta v_2 &= v_2 - v_0.
\end{aligned} \tag{7.14}$$

The (u, v) coordinates are associated with the vertices by construction and are unchanging throughout the simulation. Naturally, Δu_1, Δu_2, Δv_1, and Δv_2 will be constant too.

One important thing to keep in mind is that we are working with an approximation. We are modeling a continuous piece of cloth with a discrete set of triangles. Whereas in real life, $\mathbf{W}(u, v)$ might be any type of function over the material, we will model it as a linear function over each triangle. If the function $\mathbf{W}(u, v)$ is a linear function then the derivatives $\mathbf{W}_u, \mathbf{W}_v \in \mathbb{R}^3$ will be constant over each triangle.

Again, $\mathbf{W}(u, v)$ tells us how points are mapped and the derivatives $\mathbf{W}_u$ and $\mathbf{W}_v$ tell us how vectors are altered in that local neighborhood. In our case, $\mathbf{W}(u, v)$ is linear so the gradient is constant over the triangle surface.

We don't have an explicit expression for the deformation map $\mathbf{W}(u, v)$ but we are only looking for the gradients anyway. We know the undeformed state since this was provided with our input mesh and the deformed state is whatever state the simulation is currently in. Therefore, we can write

$$\begin{aligned} \Delta \mathbf{x}_1 &= \mathbf{W}_u \Delta u_1 + \mathbf{W}_v \Delta v_1 \\ \Delta \mathbf{x}_2 &= \mathbf{W}_u \Delta u_2 + \mathbf{W}_v \Delta v_2. \end{aligned} \tag{7.15}$$

We know all the quantities in the above equation except for $\mathbf{W}_u$ and $\mathbf{W}_v$ which we are seeking. Rewriting gives us the following solution:

$$\begin{bmatrix} \mathbf{W}_u & \mathbf{W}_v \end{bmatrix} = \begin{bmatrix} \Delta \mathbf{x}_1 & \Delta \mathbf{x}_2 \end{bmatrix} \begin{bmatrix} \Delta u_1 & \Delta u_2 \\ \Delta v_1 & \Delta v_2 \end{bmatrix}^{-1}. \tag{7.16}$$

The 2×2 matrix on the right-hand side can be precomputed since this doesn't change over time. The values for $\Delta \mathbf{x}_1$ and $\Delta \mathbf{x}_2$ can be recomputed at every time step, giving us a straightforward way to measure the stretch or compression of a triangle. Using this measure, we can formulate the following condition function:

$$\mathbf{C}(\mathbf{x}) = \begin{bmatrix} C_u(\mathbf{x}) \\ C_v(\mathbf{x}) \end{bmatrix} = a \begin{bmatrix} ||\mathbf{W}_u(\mathbf{x})|| - b_u \\ ||\mathbf{W}_v(\mathbf{x})|| - b_v \end{bmatrix}. \tag{7.17}$$

In the above equation, a is the triangle's area in the (u, v) space:

$$a = \frac{1}{2} \left\| \begin{bmatrix} \Delta u_1 \\ \Delta v_1 \\ 0 \end{bmatrix} \times \begin{bmatrix} \Delta u_2 \\ \Delta v_2 \\ 0 \end{bmatrix} \right\|. \tag{7.18}$$

Remember, (u, v) coordinates are fixed so this doesn't change over time.

There are two additional dials built into this condition function. The scalars b_u and b_v give us the ability to change the desired stretch amount, deviating from the rest pose. One trick that's often used in creating stylized animations is called (u, v) scaling. This can be done by changing the values b_u and b_v throughout the simulation.

Now, of course, this is non-physical since it makes the rest configuration of the cloth stretch or shrink, creating or removing mass from our simulated world. It is, however, very handy when creating cartoons with let's say, very stretchy limbs and you want the clothes to cover the entire body when they stretch and shrink. You can just make the clothing grow or shrink with the animated character.

Note that we only have dials that allows us to scale in the u and v direction. Typically, when the tailors are creating the garments for the digital actors, they will align the (u, v)'s of the garment with the u and v axis in a way that makes sense for the growth and shrink directions of the clothing. A good choice for the u and v direction might be parallel to the warp and weft directions of the weave.

For instance, if a sleeve of a shirt is positioned so that it is parallel to the u direction, scaling the u component will only make it longer or shorter in the length of the sleeve and not the width. On the other hand, scaling the v component will leave the length unchanged but alters the width.

Now is a good time to actually start computing the stretch forces and their derivatives. We will show the full derivation for $\mathbf{W}_u$. The final results for stretch in v direction $\mathbf{W}_v$ are derived in exactly the same way. We know the following holds for a 2×2 dimensional matrix

$$\begin{bmatrix} a & b \\ c & d \end{bmatrix}^{-1} = \frac{1}{ad - bc} \begin{bmatrix} d & -b \\ -c & a \end{bmatrix}. \tag{7.19}$$

Using this identity, in combination with Equation (7.16) and by grouping $D = \Delta u_1 \Delta v_2 - \Delta u_2 \Delta v_1$, we find the vector $\mathbf{W}_u \in \mathbb{R}^3$:

$$\mathbf{W}_u(\mathbf{x}) = \begin{bmatrix} W_{u_x}(\mathbf{x}) \\ W_{u_y}(\mathbf{x}) \\ W_{u_z}(\mathbf{x}) \end{bmatrix} = \frac{1}{D}\Big((\mathbf{x}_1 - \mathbf{x}_0)\,\Delta v_2 - (\mathbf{x}_2 - \mathbf{x}_0)\,\Delta v_1 \Big)$$

$$= \frac{1}{D} \begin{bmatrix} ((x_{1_x} - x_{0_x})\,\Delta v_2 - (x_{2_x} - x_{0_x})\,\Delta v_1) \\ ((x_{1_y} - x_{0_y})\,\Delta v_2 - (x_{2_y} - x_{0_y})\,\Delta v_1) \\ ((x_{1_z} - x_{0_z})\,\Delta v_2 - (x_{2_z} - x_{0_z})\,\Delta v_1) \end{bmatrix}.$$

$$(7.20)$$

Remember, forces are computed according to Equation (7.2). Plugging in the condition for stretch, we see that we'll need $\dfrac{\partial C_u(\mathbf{x})}{\partial \mathbf{x}_i} \in \mathbb{R}^3$

$$\frac{\partial C_u(\mathbf{x})}{\partial \mathbf{x}_i} = \frac{\partial a\,(\|\mathbf{W}_u(\mathbf{x})\| - b_u)}{\partial \mathbf{x}_i}$$

$$= a\,\frac{\partial \mathbf{W}_u(\mathbf{x})}{\partial \mathbf{x}_i}\,\hat{\mathbf{W}}_u(\mathbf{x}).$$

$$(7.21)$$

Nearly there! Now we still have to figure out what $\dfrac{\partial \mathbf{W}_u(\mathbf{x})}{\partial \mathbf{x}_i} \in \mathbb{R}^{3\times 3}$ is and we can finally compute the forces. The full matrix is computed as

$$\frac{\partial \mathbf{W}_u(\mathbf{x})}{\partial \mathbf{x}_i} = \begin{bmatrix} \dfrac{\partial W_{u_x}(\mathbf{x})}{\partial x_{i_x}} & \dfrac{\partial W_{u_x}(\mathbf{x})}{\partial x_{i_y}} & \dfrac{\partial W_{u_x}(\mathbf{x})}{\partial x_{i_z}} \\[2mm] \dfrac{\partial W_{u_y}(\mathbf{x})}{\partial x_{i_x}} & \dfrac{\partial W_{u_y}(\mathbf{x})}{\partial x_{i_y}} & \dfrac{\partial W_{u_y}(\mathbf{x})}{\partial x_{i_z}} \\[2mm] \dfrac{\partial W_{u_z}(\mathbf{x})}{\partial x_{i_x}} & \dfrac{\partial W_{u_z}(\mathbf{x})}{\partial x_{i_y}} & \dfrac{\partial W_{u_z}(\mathbf{x})}{\partial x_{i_z}} \end{bmatrix}.$$

$$(7.22)$$

Specifically, this means

$$
\frac{\partial \mathbf{W}_u(\mathbf{x})}{\partial \mathbf{x}_0} =
\begin{bmatrix}
\dfrac{\Delta v_1 - \Delta v_2}{D} & 0 & 0 \\[2.5ex]
0 & \dfrac{\Delta v_1 - \Delta v_2}{D} & 0 \\[2.5ex]
0 & 0 & \dfrac{\Delta v_1 - \Delta v_2}{D}
\end{bmatrix}
= \frac{\Delta v_1 - \Delta v_2}{D} \mathbf{I}
$$

$$
\frac{\partial \mathbf{W}_u(\mathbf{x})}{\partial \mathbf{x}_1} =
\begin{bmatrix}
\dfrac{\Delta v_2}{D} & 0 & 0 \\[2.5ex]
0 & \dfrac{\Delta v_2}{D} & 0 \\[2.5ex]
0 & 0 & \dfrac{\Delta v_2}{D}
\end{bmatrix}
= \frac{\Delta v_2}{D} \mathbf{I}
\qquad (7.23)
$$

$$
\frac{\partial \mathbf{W}_u(\mathbf{x})}{\partial \mathbf{x}_2} =
\begin{bmatrix}
\dfrac{-\Delta v_1}{D} & 0 & 0 \\[2.5ex]
0 & \dfrac{-\Delta v_1}{D} & 0 \\[2.5ex]
0 & 0 & \dfrac{-\Delta v_1}{D}
\end{bmatrix}
= \frac{-\Delta v_1}{D} \mathbf{I},
$$

with $\mathbf{I}$ the 3×3 dimensional identity matrix. Doing the same derivation for the vector $\mathbf{W}_v(\mathbf{x})$, we find

$$
\frac{\partial C_v(\mathbf{x})}{\partial \mathbf{x}_i} = \frac{\partial a \left(\|\mathbf{W}_v(\mathbf{x})\| - b_v \right)}{\partial \mathbf{x}_i} = a \frac{\partial \mathbf{W}_v(\mathbf{x})}{\partial \mathbf{x}_i} \hat{\mathbf{W}}_v(\mathbf{x})
\qquad (7.24)
$$

and

$$\frac{\partial \mathbf{W}_v(\mathbf{x})}{\partial \mathbf{x}_0} = \frac{\Delta u_2 - \Delta u_1}{D}\mathbf{I}$$

$$\frac{\partial \mathbf{W}_v(\mathbf{x})}{\partial \mathbf{x}_1} = \frac{-\Delta u_2}{D}\mathbf{I} \qquad (7.25)$$

$$\frac{\partial \mathbf{W}_v(\mathbf{x})}{\partial \mathbf{x}_2} = \frac{\Delta u_1}{D}\mathbf{I}.$$

Finally, the second derivatives can be computed using the following equalities:

$$\frac{\partial C_u^2(\mathbf{x})}{\partial \mathbf{x}_i \partial \mathbf{x}_j} = \frac{\partial \left(a \left(\frac{\partial \mathbf{W}_u(\mathbf{x})}{\partial \mathbf{x}_i} \frac{\mathbf{W}_u(\mathbf{x})}{\|\mathbf{W}_u(\mathbf{x})\|} \right) \right)}{\partial \mathbf{x}_j}$$

$$= \frac{a}{\|\mathbf{W}_u\|} \frac{\partial \mathbf{W}_u}{\partial \mathbf{x}_i} \frac{\partial \mathbf{W}_u}{\partial \mathbf{x}_j} \left(\mathbf{I} - \hat{\mathbf{W}}_u \hat{\mathbf{W}}_u^T \right) \qquad (7.26)$$

and analogously

$$\frac{\partial C_v^2(\mathbf{x})}{\partial \mathbf{x}_i \partial \mathbf{x}_j} = \frac{a}{\|\mathbf{W}_v\|} \frac{\partial \mathbf{W}_v}{\partial \mathbf{x}_i} \frac{\partial \mathbf{W}_v}{\partial \mathbf{x}_j} \left(\mathbf{I} - \hat{\mathbf{W}}_v \hat{\mathbf{W}}_v^T \right), \qquad (7.27)$$

where we found the above result using the chain rule and the equality from Equation (A.4).

7.5   SHEAR FORCES

Shearing of cloth is visualized in Figure 7.5. The shear angle can be approximated by the dot product of the deformation gradients for the u and v direction. Using this information, we can

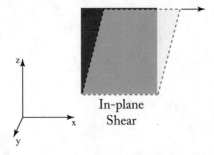

Figure 7.5: Visualization of in plane shearing for two triangles.

formulate the condition function for shearing as follows:

$$C(\mathbf{x}) = a\mathbf{W}_u^T(\mathbf{x})\mathbf{W}_v(\mathbf{x})$$

$$= a\begin{bmatrix} W_{u_x}(\mathbf{x}) & W_{u_y}(\mathbf{x}) & W_{u_z}(\mathbf{x}) \end{bmatrix} \begin{bmatrix} W_{v_x}(\mathbf{x}) \\ W_{v_y}(\mathbf{x}) \\ W_{v_z}(\mathbf{x}) \end{bmatrix} \tag{7.28}$$

$$= a\Big(W_{u_x}(\mathbf{x})W_{v_x}(\mathbf{x}) + W_{u_y}(\mathbf{x})W_{v_y}(\mathbf{x}) + W_{u_z}(\mathbf{x})W_{v_z}(\mathbf{x})\Big).$$

Just like before, we compute all partial derivatives in order to compute the forces and their derivatives. Using the definition of $\mathbf{W}_u(\mathbf{x})$, see Equation (7.20) for a reminder. The x, y, and z components of $\mathbf{W}_u(\mathbf{x})$ and $\mathbf{W}_v(\mathbf{x})$ only depend on x, y, and z components of the positions making derivates of the x component with respect to y or z equal to zero, and so on. We find

$$\frac{\partial C(\mathbf{x})}{\partial \mathbf{x}_i} = a\begin{bmatrix} \dfrac{\partial W_{u_x}(\mathbf{x})}{\partial x_{i_x}}W_{v_x} + W_{u_x}\dfrac{\partial W_{v_x}(\mathbf{x})}{\partial x_{i_x}} \\[2ex] \dfrac{\partial W_{u_y}(\mathbf{x})}{\partial x_{i_y}}W_{v_y} + W_{u_y}\dfrac{\partial W_{v_y}(\mathbf{x})}{\partial x_{i_y}} \\[2ex] \dfrac{\partial W_{u_z}(\mathbf{x})}{\partial x_{i_z}}W_{v_z} + W_{u_z}\dfrac{\partial W_{v_z}(\mathbf{x})}{\partial x_{i_z}} \end{bmatrix}. \tag{7.29}$$

While we were deriving the derivatives for the stretch condition in Equations (7.22)–(7.23), we found that the following partial derivatives were equal:

$$\frac{\partial W_{u_x}(\mathbf{x})}{\partial x_{i_x}} = \frac{\partial W_{u_y}(\mathbf{x})}{\partial x_{i_y}} = \frac{\partial W_{u_z}(\mathbf{x})}{\partial x_{i_z}}$$

$$\frac{\partial W_{v_x}(\mathbf{x})}{\partial x_{i_x}} = \frac{\partial W_{v_y}(\mathbf{x})}{\partial x_{i_y}} = \frac{\partial W_{v_z}(\mathbf{x})}{\partial x_{i_z}}.$$

(7.30)

Using this information, we can compactly write the equation as

$$\frac{\partial C(\mathbf{x})}{\partial \mathbf{x}_i} = a \left(\frac{\partial W_{u_x}(\mathbf{x})}{\partial x_{i_x}} \mathbf{W}_v + \mathbf{W}_u \frac{\partial W_{v_x}(\mathbf{x})}{\partial x_{i_x}} \right).$$

(7.31)

The second derivatives are derived in the same way, resulting in

$$\frac{\partial^2 C(\mathbf{x})}{\partial x_i x_j} = \left(\frac{\partial W_{u_x}(\mathbf{x})}{\partial x_{i_x}} \frac{\partial W_{v_x}(\mathbf{x})}{\partial x_{j_x}} + \frac{\partial W_{u_x}(\mathbf{x})}{\partial x_{j_x}} \frac{\partial W_{v_x}(\mathbf{x})}{\partial x_{i_x}} \right) \mathbf{I},$$

(7.32)

again, with $\mathbf{I}$ the 3×3 identity matrix. The second derivative is thus found as the identity matrix multiplied by a scalar.

7.6   BEND FORCES

Bend forces are defined on the angle between two triangles that share a common edge. (See Figure 7.6 for a schematic of this configuration.) The condition function states that the triangles will have minimal bend energy when the angle θ between both triangles is equal to the rest bend angle θ_0. The condition function is given by

$$\mathbf{C}(\mathbf{x}) = \theta(\mathbf{x}) - \theta_0.$$

(7.33)

We need to express θ as a function of the particle positions $\mathbf{x}$ so that we can compute the gradient. The trick we use for this is the same as documented by Pritchard [2006] and it is the following

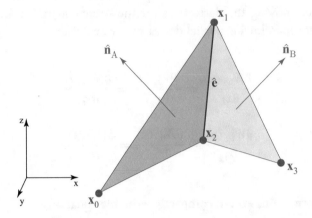

Figure 7.6: Visualization of the configuration for computing the bend forces between two neighboring triangles with common edge $\hat{e}$.

equality in trigonometry:

$$\theta(\mathbf{x}) = \arctan\left(\frac{\sin(\theta(\mathbf{x}))}{\cos(\theta(\mathbf{x}))}\right). \tag{7.34}$$

To compute the forces we will need to take the derivatives of the condition function with respect to the particle positions. Using the chain rule, and using the substitution $f(\mathbf{x}) = \dfrac{g(\mathbf{x})}{h(\mathbf{x})} = \dfrac{\sin\theta}{\cos\theta}$, we find

$$\frac{d}{d\mathbf{x}}\arctan(f(\mathbf{x})) = \frac{f'(\mathbf{x})}{f(\mathbf{x})^2 + 1} = \frac{\frac{g'(\mathbf{x})h(\mathbf{x}) - g(\mathbf{x})h'(\mathbf{x})}{h^2(\mathbf{x})}}{f(\mathbf{x})^2 + 1}$$

$$= \frac{\cos\theta\frac{\partial\sin\theta}{\partial\mathbf{x}} - \sin\theta\frac{\partial\cos\theta}{\partial\mathbf{x}}}{\cos^2\theta + \sin^2\theta} \tag{7.35}$$

$$= \cos\theta\frac{\partial\sin\theta}{\partial\mathbf{x}} - \sin\theta\frac{\partial\cos\theta}{\partial\mathbf{x}}.$$

If we can express $\cos(\theta)$ and $\sin(\theta)$ as functions of the particle positions then we can take the partial derivatives. Well, who would have thought, it turns out we can!

Let's first define some intermediate quantities. The first triangle consist of particles x_0, x_1, and x_2. The second triangle is made up of particles x_1, x_2, and x_3; see Figure 7.6. The neighboring triangles share a common edge $e(x) = x_1 - x_2$. The triangle normals are computed as

$$
\begin{aligned}
n_A(x) &= (x_2 - x_0) \times (x_1 - x_0) \\
n_B(x) &= (x_1 - x_3) \times (x_2 - x_3),
\end{aligned}
\tag{7.36}
$$

where n_A and n_B are the normals of the first and second triangle, respectively. It will be more convenient to work with the normalized vectors denoted by $\hat{n}_A$, $\hat{n}_B$, and $\hat{e}$.

We now have everything we need in order to compute the sine and cosine of the angle between the triangles based on the vertex positions:

$$
\begin{aligned}
\cos\theta &= \hat{n}_A(x) \cdot \hat{n}_B(x) \\
\sin\theta &= (\hat{n}_A(x) \times \hat{n}_B(x)) \cdot \hat{e}(x).
\end{aligned}
\tag{7.37}
$$

Just like for the stretch and shear forces, we can perform all the derivations and compute the forces and their derivatives. This is left as an exercise for the reader. A good derivation can be found in the work of Tamstorf and Grinspun [2013].

7.7   CONCLUSION

This chapter discussed the seminal work by Baraff and Witkin [1998]. We talked about how we could define internal cloth forces over triangles instead of between point masses. The model enables local anisotropic stretch or compression and offers a unified treatment of damping forces. The energies are defined based on condition functions imposed on the triangles of the cloth.

The derivations of the force derivatives for the implicit solver become a little bit more involved but we obtain simulations for which the material parameters are less dependent on the cloth geometry. This makes it much easier to model garments that have physical behavior that can be tuned much more intuitively compared to mass-spring systems. Not only that, it also allows for the matching of real-world measurements with simulations, as shown by Wang et al. [2011].

CHAPTER 8

Controlling Cloth Simulations

"$F = da$, **Director Approval**"

David Eberle

Twist on Newton's second law to indicate the high level of art direction in feature film production.

Beyond the Basics

This is definitely a somewhat more advanced topic but it's just too interesting not to mention it.

8.1 INTRODUCTION

Wojtan et al. [2006] discovered that cloth simulations can be controlled to reach certain reference positions or velocities at some time or multiple times in the simulation. As a computer graphics enthusiast, this should get you excited!

Most of the time, we're not just interested in realistically simulating garments. We probably want to have some control over the result so that we can express our creative vision. This is definitely the case for highly art directed animated movies where we don't want to just truthfully recreate physics but also want to have fine-grained control over the final look.

Let's say you're directing an animated movie and you want the clothes to look a certain way. For instance, you want the garment to have a very specific silhouette at one crucial instant in time. You can't really rely on the simulator to automatically give you what you want.

One way to do this might be to add specific wind forces over time that hopefully will blow the cloth in exactly the right shape at exactly the right time. But, this seems kind of hard too. If it is ever going to work at all, at the very least, it would require a lot of trial-and-error involving long computations for every trial run. This is a very frustrating workflow for artists.

Why not just let the computer find these forces for us? That's exactly what the method described in this chapter will do. It will find a sequence of forces that gently or not so gently, depending on what you ask it to do, blows on the cloth so that it will do precisely what you want while still being a simulated garment.

Other approaches to controlling simulations can be found in the work of McNamara et al. [2004], Kondo et al. [2005], Li et al. [2013], and Stuyck and Dutré [2016].

8.2 CONTROL PROBLEM FORMULATION

More formally, the procedure aims to find the optimal sequence of forces $\mathbf{u}$ in the N time steps leading up to the final keyframe that minimizes a goal function ϕ. We refer to N as the *control horizon*. We are trying to find an individual force for every particle in the cloth geometry and we would like to have a force for every single step in the control horizon.

8.2.1 THE GOAL FUNCTION

This goal function somehow encodes the proximity of the simulation state to the keyframes that you want the simulation to reach at prescribed times. We will see that the function also has a term that penalizes excessive forces. This is because using very strong control forces won't look natural.

The goal function $\phi(\mathbf{u}, \mathbf{Q}) \in \mathbb{R}$ looks at the control force sequence $\mathbf{u}$ and all the particle states $\mathbf{Q}$ in the control horizon N. The function expresses the difference between the simulation state $\mathbf{q_n}$ and the desired keyframe state $\mathbf{q_n^*}$ at time n as a sum over all time steps N. Intuitively, the scalar output value of this function is a measure of how far the simulation is from the desired keyframes. Of course, for controlled simulation, lower is better. Mathematically, the function is formulated as follows:

$$\phi(\mathbf{u}, \mathbf{Q}) = \frac{1}{2} \sum_{n=0}^{N} \left(||\mathbf{W_n}(\mathbf{q_n} - \mathbf{q_n^*})||^2 + \alpha_n ||\mathbf{u_n}||^2 \right), \tag{8.1}$$

where $\mathbf{W}_n$ and α_n are weights that can be tuned by the artist to control the simulated behavior. These parameters will be discussed more thoroughly later in the text. The matrix $\mathbf{Q}$ contains all particle states over all time steps. The vector $\mathbf{u}$ is the concatenation of all control forces applied over the entire control horizon. This is a long vector. For P particles over N time steps this will be $\mathbf{u} \in \mathbb{R}^{3PN}$:

$$\mathbf{u} = \begin{bmatrix} \mathbf{u}_{0,0}, & \mathbf{u}_{1,0}, & \dots, & \mathbf{u}_{P-1,0}, & \mathbf{u}_{0,1}, & \dots, & \mathbf{u}_{P-1,n-1} \end{bmatrix} \tag{8.2}$$

with $\mathbf{u}_{p,n} \in \mathbb{R}^3$ the force on particle p at time step n.

The reference state $\mathbf{q}_n^* \in \mathbb{R}^{6P}$ consists of all P particle positions and velocities at frame n. Note that this is only when we have a keyframe at time n. Otherwise, there will be no contribution in the sum in the first term of the goal function.

Let's have a closer look at the intuitive meaning of the goal function given in Equation (8.1). Minimizing the first term in the sum will match the cloth particles with the keyframe state. This is the part that will make sure that our particles are controlled to reach the goal states at the right time.

Okay sounds great, so why do we need a second part in the goal function? Turns out, there is a very good reason! The sum over the norm of the forces will make sure that the applied control forces won't get too big. When this term gets too large, it will dominate the value of the goal function and it will be in the best interest of the optimizer to lower the strength of the control forces since this would immediately lower the cost of the goal function.

If the control forces get too big, it might be easy to hit the goal states but the force will be so excessive that the results will look very unnatural and forced. The second part of the sum acts as a regularization term to discourage from using overly strong control forces $\mathbf{u}$. As such, preventing unnatural hand-of-god-like simulation results.

8.2.2 TUNING THE GOAL FUNCTION

Now that we have a goal function defined, we would like to have some artistic control over the end result. We would like to tune the trade-off between the following two extremes.

- **Case 1.** Applying a lot of control in order to make sure that we hit the keyframes well. This will make the particle states reach the goal states but it will probably do so in a forced and unnatural looking way.

- **Case 2.** Applying a gentle amount of force where the particle states don't quite reach the keyframes precisely. This won't give you exact control but it will make the controlled results look much more natural.

The trade-off between the above two cases can be tuned using the goal function weights $\mathbf{W}_n \in \mathbb{R}^{6P \times 6P}$ and $\alpha_n \in \mathbb{R}$. The force regularizer α_n will penalize the use of excessive force. Having a small weight for α_n compared to $\mathbf{W}_n$ will result in the first case. Having a big weight α_n is described in the second case. An ideal value will likely be somewhere in between these two extreme cases.

The matrix $\mathbf{W}_n$ gives a weight to the particle state at time step n. This matrix can be constructed to add importance to the individual positions and velocities of the individual particles at time n. They can vary for different keyframes at different steps.

We should also point out that is very likely that you won't have a keyframe state for every time step in the control horizon. Typically, there will only be a few keyframe shapes that you want the simulation to hit. You will want the optimizer to find a physically plausible path between these states. In essence, doing a physically based interpolation using simulation. This means that $\mathbf{W}_n$ will only be non-zero for time steps that have a corresponding keyframe. Because otherwise, the term will just be dropped from the sum.

8.2.3   MINIMIZING THE GOAL FUNCTION

We have formulated a goal function that we want to minimize. If we can find the control sequence that minimizes this function, then we have controlled simulations. It's that easy! Well, it turns out that minimizing such a complex goal function which depends on a simulation result isn't the easiest thing ever.

One way to perform this optimization is to use gradient descent. We iteratively take a step in the gradient direction, hoping that every step will make the goal function a little bit smaller. We keep doing this until we converge and reach the minimum. We point the interested reader to the excellent book by Nocedal and Wright [1999].

That sounds simple! True, but we also want to do this in a reasonable amount of time. We will see that computing these gradients naively just takes way too much computation time. Another big issue is that the goal function is not guaranteed to be a nice and smooth convex function with monotonic convergence. This means that gradient descent might fail us. Overcoming this is not straightforward, so we won't be discussing it here any further.

Let's have a look at how this gradient of the goal function with respect to all control forces $\frac{d\phi}{d\mathbf{u}}$ could be computed in the naive way. Taking the derivative, we find

$$\frac{d\phi}{d\mathbf{u}} = \frac{\partial \phi}{\partial \mathbf{Q}} \frac{d\mathbf{Q}}{d\mathbf{u}} + \frac{\partial \phi}{\partial \mathbf{u}}. \tag{8.3}$$

However, this is *much* too costly given the computation of the full $\frac{d\mathbf{Q}}{d\mathbf{u}}$ matrix. The matrix $\mathbf{Q}$ contains the full simulation state for all particles over all time steps. Instead, we compute these gradients by reformulating the problem in terms of the adjoint states $\hat{\mathbf{Q}}$. Please have a look at the excellent explanation by Wojtan et al. [2006] to see how these formulas are derived. A good overview can also be found in the work of McNamara et al. [2004] who used the adjoint method to control fluid simulations. The adjoint formulation of the goal function gradient is given by

$$\frac{d\phi}{d\mathbf{u}} = \hat{Q}^T \frac{\partial \mathbf{F}}{\partial \mathbf{u}} + \frac{\partial \phi}{\partial \mathbf{u}}. \tag{8.4}$$

In the equation above, $\mathbf{F}(\mathbf{Q}, \mathbf{u})$ encapsulates the time step formulae between the states $\mathbf{Q}$ at different times. We will clarify this later in the chapter. For now, it describes how particles get updated from one time step to the next. We see that we no longer need the excessively expensive $\frac{d\mathbf{Q}}{d\mathbf{u}}$ term in order to compute the derivative. If we can find these so-called adjoint states $\hat{\mathbf{Q}}$, we can use Equation (8.4) to obtain the gradient in an efficient way.

The overall optimization is achieved using an iterative two-step process. In every iteration, we compute a gradient that we can use to update our best estimate of the control force sequence. We start the method by initializing the control forces. This could be anything you want. The most simple approach is to initialize them to just be zero forces. Something more elaborate would be where we use a different strategy to get an estimate of the control forces. We can then use those as an initial guess. These are then refined using gradient descent optimization.

To summarize, once we have a control force vector to start with, we perform the following two steps in a loop until we converge or decide that we've spent enough time on this problem.

1. Gradient computation starts by running a standard forward cloth simulation and by applying the current best guess of the sequence of optimal forces. We will refer to this step as the *forward simulation*.

2. The second step of the gradient computation consists of a simulation backward in time where the adjoint states are computed. Once all adjoint states have been computed over the optimized frames, the states are mathematically mapped to the gradient. We refer to this step as the *backward simulation*.

8.3   ADJOINT STATE COMPUTATION

In this section, we will have a look at how these adjoint states are actually computed. The adjoint states $\hat{\mathbf{q}}_n = \langle \hat{\mathbf{x}}_n, \hat{\mathbf{v}}_n \rangle$ at time n are computed using the following equation:

$$\hat{\mathbf{Q}}_n = \left(\frac{\partial \mathbf{F}}{\partial \mathbf{Q}}\right)^T \hat{\mathbf{Q}}_{n+1} + \left(\frac{\partial \phi}{\partial \mathbf{Q}}\right)^T . \tag{8.5}$$

The derivation of this formula can be found in Wojtan et al. [2006]. Here, we'll just assume the author wasn't lying and accept this as the one true formula. Note that in this chapter, $\hat{\mathbf{q}}_n$ refers to the adjoint state and not the normalized vector. Once computed for all time steps, these adjoint states are then mapped to the gradient using the equation given in (8.4).

One funny thing about this formulation is that the prior adjoint state depends on the next one. We will have to run a simulation backward in time in order to solve this for all time steps. We start by initializing the final adjoint state and then work our way back to the beginning of the simulation. This is why we named this phase the backward simulation step.

The equation given in (8.5) is still a little bit vague. What is $\frac{\partial \mathbf{F}}{\partial \mathbf{Q}}$ supposed to be? Let's look into it some more here. We mentioned earlier that $\mathbf{F}$ is what takes the particle states from one time step to the next. In our standard forward simulation, we used the linearized backward Euler integration scheme to accomplish this. Mathematically, computing the adjoint states based on the linearized scheme is the right thing to do in order to compute the correct gradient. However, Wojtan et al. [2006] states that doing so will lead to a dimensional explosion in the derivatives which again would make the method computationally untracktable.

This issue can be overcome by computing the adjoint states corresponding to the backward Euler scheme instead of its linearized version. This means that we're no longer computing the exact gradients for our simulations. However, we can compute a gradient that is very similar at a much cheaper cost. For computer graphics purposes, this is definitely worth the trade-off. Recall that the backward Euler scheme is given by

$$\mathbf{q}_{n+1} = \mathbf{q}_n + h\mathbf{V}(\mathbf{q}_{n+1}) \tag{8.6}$$

with $\dfrac{d\mathbf{q}}{dt} = \mathbf{V}(\mathbf{q})$. Remember that h was the time duration with which we advance the simulation in a single simulation step. If we substitute this into Equation (8.5) for computing the adjoint states, we find

$$\hat{\mathbf{q}}_n = \hat{\mathbf{q}}_{n+1} + h \left(\left. \frac{\partial \mathbf{V}}{\partial \mathbf{q}} \right|_n \right)^T \hat{\mathbf{q}}_n + \left(\frac{\partial \phi}{\partial \mathbf{q}_n} \right)^T. \tag{8.7}$$

This adjoint state computation is linear because the Jacobian $\left. \dfrac{\partial \mathbf{V}}{\partial \mathbf{q}} \right|_n$ is known from the corresponding time step in the forward simulation. So, if we apply the following backward Euler scheme to Equation (8.7)

$$\begin{aligned} \mathbf{v}_{n+1} &= \mathbf{v}_n + h\mathbf{M}^{-1}\mathbf{f}_{n+1} \\ \mathbf{x}_{n+1} &= \mathbf{x}_n + h\mathbf{v}_{n+1}, \end{aligned} \tag{8.8}$$

we get

$$\hat{\mathbf{v}}_n = \hat{\mathbf{v}}_{n+1} + h\mathbf{M}^{-1} \left(\left. \frac{\partial \mathbf{f}}{\partial \mathbf{v}} \right|_n \right)^T \hat{\mathbf{v}}_n + h\hat{\mathbf{x}}_n + \left(\frac{\partial \phi}{\partial \mathbf{v}_n} \right)^T$$

$$\hat{\mathbf{x}}_n = \hat{\mathbf{x}}_{n+1} + h\mathbf{M}^{-1} \left(\left. \frac{\partial \mathbf{f}}{\partial \mathbf{x}} \right|_n \right)^T \hat{\mathbf{v}}_n + \left(\frac{\partial \phi}{\partial \mathbf{x}_n} \right)^T. \tag{8.9}$$

After grouping terms together, we find that, in the backward simulation, we will have to solve a system of the form $\mathbf{Ax} = \mathbf{b}$ in order to compute the adjoint velocities $\hat{\mathbf{v}}_n$. The system is as follows:

$$
\left(\mathbf{M} - h \left(\frac{\partial \mathbf{f}}{\partial \mathbf{v}} \right)^T - h^2 \left(\frac{\partial \mathbf{f}}{\partial \mathbf{x}} \right)^T \right) \hat{\mathbf{v}}_n =
$$
$$
\mathbf{M} \left(\hat{\mathbf{v}}_{n+1} + \left(\frac{\partial \phi}{\partial \mathbf{v}_n} \right)^T + h \hat{\mathbf{x}}_{n+1} + h \left(\frac{\partial \phi}{\partial \mathbf{x}_n} \right)^T \right).
$$

(8.10)

The right-hand side contains all known quantities. This means that it can just be computed at every time step. One remarkable thing about this adjoint computation is that the left-hand side of this system is exactly the same one as used in the forward simulation at the corresponding time step. If you're not sure about this, go ahead and look back to the chapter on implicit integration. We can just save this matrix at every time step in the forward simulation and then use it again for the backward simulation.

Not that much extra work. Just a little bit of extra storage. Well, by now you should know that these matrices are big and having to save one for every time step will require a significant amount of computer memory. This too will have a significant impact on performance. This trade-off between less computation requirements but more storage needs is typical for adjoint methods.

Before starting a backward simulation, we will have to initialize the final adjoint states. Initialization is done using

$$
\hat{\mathbf{q}}_N = \left(\frac{\partial \phi}{\partial \mathbf{q}_N} \right)^T.
$$

(8.11)

These states $\frac{\partial \phi}{\partial \mathbf{x}_N}$ and $\frac{\partial \phi}{\partial \mathbf{v}_N}$ equal $\mathbf{x}_N - \mathbf{x}_N^*$ and $\mathbf{v}_N - \mathbf{v}_N^*$ multiplied by their respective goal function weights.

After solving this system, the adjoint velocities $\hat{\mathbf{v}}_n$ are known and we can compute the adjoint positions $\hat{\mathbf{x}}_n$ using Equation (8.9). Again, we use the same Jacobian matrix $\frac{\partial \mathbf{f}}{\partial \mathbf{x}}$ that we saved in the corresponding step in the forward simulation.

8.4   UPDATING CONTROL FORCES

After solving the system to obtain the adjoint velocities, the adjoint positions can easily be computed using Equation (8.9). Given these adjoint states, the gradient vector is computed using the Formula (8.4).

In Wojtan et al. [2006], a force is computed for every single particle for every single time step. Typically, a simulation has to take multiple time steps in order to advance one frame. In our experiments, we have found that applying the same control force per particle over all time steps needed to advance the simulation one frame produces much better results.

In theory, this would make the approach less expressive. But, we have found that the reduced dimensionality of the control space significantly outweighs this reduced expressive power because of faster convergence and smoother results. This is achieved by accumulating the contributions of all the sub-steps to the corresponding frame.

The control forces are applied to the particles and explicitly integrated into the simulation. Only considering the contribution of the control force, we have

$$\mathbf{v}_n = \mathbf{v}_n + h\mathbf{M}^{-1}\mathbf{u}_n. \tag{8.12}$$

For a single cloth particle, $\frac{\partial \mathbf{F}}{\partial \mathbf{u}} \in \mathbb{R}^{6P \times 3P}$ maps the $\mathbb{R}^{3P}$ control vector back to the $\mathbb{R}^{6P}$ state space. For a single particle p we have

$$\frac{\partial \mathbf{F}}{\partial \mathbf{u}} = \begin{bmatrix} 0 & 0 & 0 \\ 0 & 0 & 0 \\ 0 & 0 & 0 \\ \frac{h}{m_p} & 0 & 0 \\ 0 & \frac{h}{m_p} & 0 \\ 0 & 0 & \frac{h}{m_p} \end{bmatrix}, \tag{8.13}$$

with m_p the mass of particle p on which the force is being applied to. Finally, the formula that computes the gradient for particle p is thus given by

$$\frac{d\phi}{d\mathbf{u}_{p,n}} = \frac{h}{m_p}\hat{\mathbf{v}}_{p,n} + \alpha_n\frac{h}{m_p}\mathbf{u}_{p,n}. \tag{8.14}$$

The negative gradient can then be used to obtain an updated estimate of the control forces. The Broyden–Fletcher–Goldfarb–Shanno (BFGS) algorithm or line search can be used to find a good step size δ. We refer to Nocedal and Wright [1999] for an in-depth explanation of these techniques. This step size computation is needed since the gradient just points in the direction of steepest ascent. We still have to figure out how far along this direction we'd like our update to be. When the step size is too small we might not be making enough improvement with a single iteration. In contrast, when the step is too large, we might overshoot our goal and the new solution will actually be worse! The BFGS algorithm is a quasi-Newton method that tries to estimate the second derivative based solely on first derivative information. A good estimate of the second derivative can be used to drastically improve the convergence rate of the algorithm without having to actually compute the Hessian matrix.

Given an adequate step size, an improved optimal control force sequence is then obtained by performing a gradient descent step

$$\mathbf{u} = \mathbf{u} - \delta \frac{d\phi}{d\mathbf{u}}. \tag{8.15}$$

Since the negative gradient is used in gradient descent, we can clearly see the effect of the control force weight scalar α_n by combining (8.14) and (8.15). This value is effectively the fraction of the current force sequence that will be subtracted from itself, scaled by the gradient descent step size δ.

8.5   CREATING KEYFRAMES

We've been talking about keyframes or reference states $\mathbf{q}^*$, but where do they come from? A keyframe should provide information about the positions or velocities, or both, that we want the cloth to have at a certain point in time. To obtain keyframe geometry, we can treat the cloth shape as a traditional triangle mesh and perform operations on it to model a new desired shape. As long as the topology doesn't change, this can immediately be used as the goal positions.

One obvious counter argument you can make about this approach, is that you're asking an artist to create a physically plausible state by hand. This is exactly what we want to avoid by using physics-based animation because it's so cumbersome to do.

Some work has been proposed on combining cloth dynamics with the geometrical modeling of shapes. Intuitively, this means that the artist can move a few points on the garment and the other vertices will follow this movement in a physically plausible way by using quasi-static simulation. For more details on sculpting simulations, we refer to the work of Stuyck [2017] and Stuyck and Dutré [2016].

It is a lot trickier to create goal velocities for the particles. There's no straightforward way to model velocities since these are only noticeable over time. In contrast, geometry modeling is something that is done at a certain point in time. There's no time aspect to it. One way to obtain goal velocities is to sculpt two shapes at two subsequent time steps and to compute the velocity for every particle that would bring the particle from the first shape to the next one.

8.6   CONCLUSION

In this chapter, we introduced a method to control cloth simulations. Instead of having the artist find the correct simulation settings through a trial-and-error approach, we showed how this problem can be formulated into a goal function that can then be minimized by applying control forces. The control problem is iteratively solved using gradient descent. We found out that naive computation of the gradient would be intractable. In order to keep the computations tractable, we made use of the adjoint method where we computed the adjoint states with respect to the backward Euler integration scheme. This differs from the linearized backward Euler scheme used in the forward simulation but it leads to much more efficient evaluations of the gradient. This gradient is approximate but this doesn't pose significant problems for typical control problems.

We use line search to find a good step length to update our control force sequence. This won't always work since the goal function isn't necessarily a smooth convex function. The algorithm is likely to get stuck in local minima which are hard to detect at runtime.

Up until now, we discussed having a control force per particle, per time step. This very quickly leads to a high-dimensional search space. Alternative approaches would be to not have a force per particle but a force over a subset of particles that is spread out over neighboring particles. This would provide less control but faster convergence. Another approach is to have wind forces spatially defined, affecting the cloth as it moves past these spatially localized wind forces.

A major issue with the approach presented in this chapter is that collision information isn't included in the gradient. The gradient might point into a direction that seems optimal to the optimizer but would result in cloth collisions making this strategy unusable. This is a difficult and unsolved problem. Despite that, the method is still very powerful given that the requested keyframes are reasonable. More information about collisions will be given in the next chapter.

CHAPTER 9

Collision Detection and Response

"OH YEAH!!!!!"

Kool-Aid Man

He can often be found answering the call of children by running through walls and furnishings.

9.1 INTRODUCTION

You might have noticed but we barely mentioned collision detection and response at all throughout this text. It is an important but tricky aspect of cloth simulation that it is often the bottleneck in modern visual effects. Collision handling is split into two phases: the **collision detection phase** and the **collision response phase**. First, we try to find colliding triangles or particles. Once we know the culprits we will try to resolve the collisions using an appropriate collision response. Two different types of collisions can be discerned.

- **Cloth-cloth** collisions happen when the garment collides with itself or another garment. For example, when wearing a shirt and sweater, you wouldn't want the shirt to pass through the sweater.

- **Object-cloth** collisions happen when the cloth collides with other objects. Think about how a shirt is constantly colliding with the body.

Collision detection is probably the easier task of the two to get working correctly. Many geometric tests exist to figure out whether triangles or particles are colliding and it is a matter of implementing these correctly. However, a lot of computation time will be spend on detecting collisions.

Collision response can be hard to get right. Applying changes to fix the collision could easily create odd-looking artifacts in the simulation. Additionally, resolving one collision can create new collisions which in turn need to get resolved as well. There's no guarantee that the algorithm ever converges! Obviously, this can be problematic and there exist techniques to deal with this situation which will be discussed later in this chapter. Another difficult situation is when cloth slides over itself or another garment. The cloth is in contact with itself so it is colliding but we do not want to restrict it too much since this would create snagging artifacts.

9.2   COLLISION DETECTION

The collision detection phase can be split into a broad phase, a mid phase, and a narrow phase. We recommend the excellent book by Ericson [2004] and the freely available chapter on collision detection in Akenine-Möller et al. [2018]. The broad phase is first performed and is meant to quickly discard object pairs that are definitely not colliding. As a second step, the mid phase will look at the overlapping primitives between object pairs. As a final step, the narrow phase then takes a closer look at primitive pairs that could potentially be close to each other. We give a quick overview.

- The **broad phase** works by looking at objects that overlap. The workhorse for the broad phase is the sweep and prune algorithm [Akenine-Möller et al., 2018]. However, for cloth simulations we typically assume a limited number of meshes in the scene and we can safely skip the broad phase and start with the mid phase.

- The **mid phase** works on pairs of objects to find primitives in the object that overlap. The phase makes use of spatial acceleration structures that can quickly prune particle pairs that definitely won't collide. This significantly reduces the number of expensive collision tests that need to be performed in the narrow phase. Possible acceleration data structures are bounding volume hierarchies, acceleration grids, or k-d trees.

- In the **narrow phase**, the remaining potential cloth-cloth intersections can then be computed using particle-triangle and edge-edge collision tests. The cloth-solid intersections are computed by checking the cloth particles with respect to the faces of the solid object.

When using axis-aligned bounding boxes for the triangles in the acceleration structures, we typically enlarge the bounding box by the thickness of the cloth, e.g., 10^{-3}m. Of course, as particles and triangles move around in the simulation the acceleration structure has to be updated in every iteration.

Collision detection algorithms can be classified by when they look for collisions. To be more precise, we have the following.

- **Discrete time** collision detection will look for all particles that are in close proximity at the beginning of the time step and will add constraints or penalty forces to particles that appear to be colliding. This will hopefully prevent the collision but there are no guarantees and this is why algorithms need failure modes to recover from collision. These methods are also known as *a posteriori*. To reliably resolve the collision continuous time algorithms will be needed.

- In contrast, **continuous time** algorithms look at the particle trajectories and will look for the instant in time that the first collision happens. The simulation will then be advanced until this time of first collision at which the collision can be resolved. This is also known as *a priori*. This is performed in an iterative fashion until all collisions are resolved.

9.2.1 BOUNDING VOLUME HIERARCHIES

Detecting a collision between two complex geometries can be very expensive, especially with ever increasing complexity of geometry. Geometry meshes with tens of thousand of triangles are pretty common nowadays. It would be extremely inefficient to simply compare every particle on the mesh with every other particle in the scene. The Bounding Volume Hierarchy (BVH) is one particular acceleration structure that can be used in the broad phase to get better efficiency and time complexity. In this text, we will focus on the BVH. For completeness, other algorithms that are frequently used for efficient pruning are uniform grids, hierarchical grids, binary space partitioning, and k-d trees.

To accelerate collision detection, complex geometries are often contained in a surrounding bounding volume. This volume can be a box, sphere, cylinder, or any other primitive that's cheap to test for intersections. Only when there's an intersection with the bounding volume do we need to intersect with the primitives inside. Another way of saying this is that only when the bounding volumes overlap could there potentially be an intersection and further investigation in the narrow phase is needed. This will save a lot of computation since large pieces of geometry can quickly be pruned since there won't be any intersections with the surrounding volume.

At the lowest level, triangles are embedded in a bounding volume. If we perform an intersection test of a single point with all the triangles then the complexity would scale linearly with the number of triangles n in the mesh. This gives the following time complexity $\mathcal{O}(n)$ using big O notation, where n is the number of triangles. We can do better than this. By grouping bounding volumes in a hierarchy; simply put, by constructing bigger bounding volumes containing many smaller volumes. The hierarchical bounding volume will have a way better time complexity of $\mathcal{O}(\log n)$.

The most commonly used bounding volume is the Axis-Aligned Bounding Box, often abbreviated as AABB. The box is aligned with the xyz-axis of the world coordinate system and it is just big enough to fully contain the geometry inside. It has the following desired properties that we are always looking for in a bounding volume.

1. It is cheap to test for intersection with the bounding volume.

2. It tightly fits around the geometry inside.

3. It is inexpensive to compute.

4. It is easy to transform.

5. It makes efficient use of computer memory.

We refer to Chapters 4 and 6 in Ericson [2004] for a thorough explanation and discussion.

9.2.2   BASIC PRIMITIVE TESTS

After pruning most of the possible intersections using the bounding volume hierarchy, we will have to perform additional test to see whether the remaining geometry is colliding or intersecting. There are a number of primitive tests available for exactly this. For example, commonly used intersection tests are

- closest point on plane to point;

- closest point on line segment to point;

- closest point on AABB to point;

- closest point on triangle to point;

- closest point of two line segments; and

- closest point of two triangles.

We refer the reader to Chapter 5 in Ericson [2004] for an in-detail explanation and example code of all the different primitive tests.

9.3   COLLISION RESPONSE

Collision response can be treated separately for cloth-cloth and object-cloth collisions. We will introduce an approach for each in the next two sections.

9.3.1   CLOTH-CLOTH COLLISION RESPONSE

A lot of cloth-cloth collisions, also named self-collisions, can be prevented by temporarily adding a strongly damped repulsive spring to particles that are about to collide. This will accelerate the particles away from each other, hopefully preventing the collision from happening. Baraff and Witkin [1998] combine damped spring forces for self-collisions and constraints for object-cloth collisions. Both are integrated in the implicit integration. The spring forces and their derivatives are added to the linear system solve for stability.

Repulsion forces are essential to keep the number of collisions tractable but we will still need to resolve some collisions that still occur by applying impulses to the particle velocities.

A more sophisticated method for cloth-cloth collision response was presented by Bridson et al. [2003]. Impulses are applied to instantly update the particle velocities to resolve the collisions. They also apply repulsive forces for when particles get too close together to prevent the majority of possible collision events. More precisely, repulsion forces are added when the particles are at a proximity similar in size to the cloth thickness.

9.3.2   OBJECT-CLOTH COLLISION RESPONSE

As before, there are numerous ways to model this type of collision response. One way is to directly alter the particle state to resolve the collision. In this approach, we simply update the violating particle's position and velocity directly so that the cloth particle is no longer colliding with the object. This works reasonably well for explicit methods but tends to create un-smooth results for implicit methods due to the lack of integration and propagation of the update to the surrounding connected cloth particles.

Bridson et al. [2002] use level sets to model collisions with objects. A grid is constructed around the simulation scene. Every grid cell will then be assigned a real number. All grid cells are initialized with a small positive value. As a second step of the initialization, all cells that lie within the collision objects are assigned a negative value. A fast marching method is used to convert this grid into a level set ϕ. This level set is then used to find collisions with the cloth.

Let's say we have a collision with a cloth particle at point $\mathbf{p}$ with velocity $\mathbf{v}_p$. A collision occurs when the level set is negative $\phi(\mathbf{p}) < 0$. We can use the level set directly to compute the normal pointing outwards of the object. The normal $\mathbf{n}$ is computed as

$$\mathbf{n} = \nabla\phi. \tag{9.1}$$

The easiest way to resolve this collision is to simply push the point outward in the direction of the normal $\nabla\phi$. Collision response often create *popping* artifacts when the cloth points are pushed outside the collision object. Therefore, the authors constructed a more elaborate scheme that is able to resolve future interferences of the cloth particle with the objects. This enhances the stability and smoothness of the results. In order to do so we will need the velocity $\mathbf{v}$ of the point on the collision object. Have a look at Bridson et al. [2002] and PhysBAM[1] to find implementation details.

9.4   DISCUSSION

Resolving one collision might create new collision so the algorithm could take a long time to converge. When this happens, a failsafe method can be activated that will treat groups of particles as rigid bodies that grow when more collisions are detected. This technique known as rigid impact zones was presented by Bridson et al. [2002] and is borrowed from rigid body dynamics. This is one way to handle the problem of the problem of potentially never-ending collisions. Another way is to run a maximum number of iterations and then try to resolve collisions later after they have occurred. This is of course not the correct solution since this can create simulations where the cloth goes through the body but this method might sufficient for many practical purposes.

[1]http://physbam.stanford.edu/

9.5   FURTHER READING

One way to find whether nearby cloth regions have interpenetrated is to use history. If we know that the cloth started in a valid configuration, we can track it over time to figure out what side cloth regions should be on. The problem with history-based algorithms is that any mistake along the way will create persistent tangles that can't be resolved. A good approach to resolving cloth-cloth collisions using a history-free cloth collision response algorithm based on global intersection analysis of cloth meshes at each simulation step is given by Baraff et al. [2003]. Such a global intersection analysis will be necessary to obtain robust simulation results in a production setting where cloth regularly gets pinched in between the elbows or armpit areas.

We refer the reader to the following resources for more information Provot [1997], Volino and Magnenat-Thalmann [2000], Bridson et al. [2002], and Schvartzman et al. [2010].

9.6   CONCLUSION

We have briefly introduced an overview to the collision detection and response problem and provided references to more in-depth explanations. It is essential in most cloth simulations due to the tight coupling of the body with the clothing. First, colliding primitives need to be detected in an efficient manner. This is typically solved in different steps, each looking at collisions at different scales. Colliding particles are modified with a collision response in order to resolve the collision. The type of response can differ based on the type of interaction. For complex scenes, discrete collision detection often doesn't suffice and continuous collision detection is needed. A robust implementation will need to be able to gracefully recover from collisions when they occur. It is not uncommon that cloth particles get pinched in armpits or elbows and this needs to be treated in a separate way.

CHAPTER 10

What's Next

> "I never look back, darling. It distracts from the now."
>
> *Edna Mode*

In this chapter, we will talk a little bit more about advanced topics and point you to further reading.

10.1 REAL-TIME APPLICATIONS

If you are interested in games and virtual reality applications, one method that's particularly suited for obtaining stable real-time results is *position-based dynamics* by Müller et al. [2007] and Macklin et al. [2016]. The technique has received widespread acceptance in the research community and industry as a fast and stable way to obtain plausible simulations. In fact, it has implementation in many state-of-the-art physics engines such as NVIDIA PhysX,[1] Havok Cloth,[2] Maya nCloth,[3] and Bullet.[4]

The method avoids needing expensive implicit integration and instead works by modifying the positions of the particles directly. The cloth behavior is described by a set of constraints that are iteratively solved in every time step. The method is not physically based but it produces visually pleasing results in a surprisingly small amount of time and is therefore commonly used in the industry. A very good tutorial was given at Eurographics 2017 by Bender et al. [2017].

Another interesting and fast approach is the use of projective dynamics by Bouaziz et al. [2014]. It bridges finite element methods and position-based dynamics. It is similar to position-based dynamics but inspired by physically based continuum mechanics. The technique has applications ranging from the simulation of deformable solids, cloth, and thin shells.

---

[1]https://developer.nvidia.com/physx-sdk
[2]https://www.havok.com/cloth/
[3]https://www.autodesk.com/products/maya/overview
[4]https://pybullet.org/wordpress/

10.2   SUBSPACE CLOTH SIMULATION

There is another very different category of approaches to real-time cloth simulations. Instead of doing a full simulation, we represent the cloth in a low-dimensional subspace. This subspace captures most of the cloth dynamics in the basis vectors. We advance the simulation in this low-dimensional space and project the garment back into the full-dimensional space.

This subspace is built once during a preprocessing phase using precomputed simulation results. To obtain good results, the subspace needs to be chosen well. The training data needs to be sufficient in number and contain adequate variations of simulated examples and poses. This approach leads to very fast computation times at runtime. The drawback is the fact that collisions aren't handled properly. Multiple subspace approaches have been proposed and we refer the reader to the work of De Aguiar et al. [2010], Kim et al. [2013], and Hahn et al. [2014].

10.3   ALTERNATIVE CLOTH MODELS

We have seen how different materials can be modeled using different spring stiffnesses for the mass-spring system presented in Chapter 4. Alternatively, a variety of materials can be modeled by controlling the stiffnesses for the forces in the continuum-inspired model presented in Chapter 7. In addition to these models, there are other cloth models that can be used to represent different materials.

For instance, knitted garments can be simulated using a yarn level simulation such as the technique proposed by Kaldor et al. [2008]. Another way to model cloth is to use the model presented by Choi and Ko [2002] where they explicitly model the post-buckling instability by assuming that cloth buckles immediately at the onset of compression. This model will make it easier for folds to persist and evolve over time making simulations more realistic and interesting. It is worthwhile modeling this buckling effect because the fully implicit integration by Baraff and Witkin [1998] introduces a lot of damping, preventing folds from persisting.

Apart from cloth, different flexible thin shell objects such as hats, leaves, and aluminum cans can be modeled using a cloth simulator. These are typically called thin shell models and require a more advanced expression for the bending energy. For a good starting point, we refer the reader to the work of Grinspun et al. [2003] on discrete shells. Another approach to preserving better wrinkles and folds in the garment is proposed by Bridson et al. [2003]. Recent work by Li et al. [2018] focuses on designing 2D sewing patterns that will create folds and pleats based on user sketches.

Another interesting phenomenon that can be simulated is the tearing and cracking of cloth and thin sheets. Pfaff et al. [2014] propose a technique where the triangle mesh is dynamically restructured to adaptively maintain detail where it is required such as along the tears. Their model allows to simulate a wide range of materials with different fracture behaviors. A method for tearing cloth with frayed edges is presented by Metaaphanon et al. [2009].

One way to speed up simulation time is to only spend significant computational effort on parts of the garment where it is really needed. For computer animations, the only thing that really matters is what is visible from the viewpoint of the camera. Geometry that is occluded from the camera does not require the same level of detail as garments that are close to and facing the camera. One way to incorporate this in your simulator is to make use of view-dependent adaptive cloth simulation such as proposed by Koh et al. [2014].

10.4 ART DIRECTING CLOTH

We saw a method to control cloth simulations in Chapter 8. Many alternative approaches to influence cloth animations have been investigated. Kondo et al. [2005] enforce trajectory constraints on a finite element-based elastic body and adapts the stiffness matrix in order to match key poses. Bergou et al. [2007] proposed a method which they named *TRACKS* for thin shell simulations where physically based details are added to a given coarse animation. In the same vein as this coarse-to-fine design cycle, Cutler et al. [2005] present a kinematic system for creating art-directed wrinkles on costumes for digital characters. Details are added as deformations based on wrinkle patterns. Bhat et al. [2003] propose a way to optimize estimated simulation parameters using simulated annealing to closely resemble a real-life video recording of cloth.

10.5 CLOTH RENDERING

We talked about a lot of topics in this book so far but haven't discussed how to visualize the simulated clothing. The easiest and most straightforward way is to just render the simulated triangles. This will create nice looking renders when the resolution is high enough, meaning that triangles are small enough to not be too easily noticed by the viewer. This will work but might suffer from noticeable angular features, resulting from the interconnection of the triangles by straight edges. For cloth, a smoother surface is usually expected.

This can be resolved by working with two separate meshes. A simulation mesh and a render mesh. The simulation mesh will drive the motion for the high resolution render mesh. Alternatively, the simulation mesh can be subdivided at render time to create higher resolution geometry. Several approaches can be found in the work of DeRose et al. [1998], Grinspun and Schröder [2001], and Bridson et al. [2002].

CHAPTER 11

Conclusions

In this book, we introduced different approaches to cloth simulation. We hope you enjoyed reading through the document and feel motivated to dive deeper into the topic of cloth simulation. We started by explaining the cloth fundamentals and how we can integrate these over time using explicit integration. It is a very simple approach but suffers from frequent instabilities unless we take very small time steps. To alleviate this restriction, we turned to implicit integration where we saw how we needed to compute the force derivatives. We discussed mass-spring systems and explained how these can be solved using an optimization reformulation.

Mass-spring models are very easy to set-up but it is very difficult to control them in order to represent real-world materials and garments. This is due to the fact that the behavior is very dependent on the interconnection of the particles with springs. To ameliorate the situation, we saw how a continuum-inspired approach to the problem can be used. The cloth is no longer discretized using point masses and springs but forces are defined over triangles as a whole. We looked at a way to control these cloth simulations given reference particle states using optimization with the help of the adjoint formulation. We finished this document by giving an overview to the collision detection and response problem. Additionally, we discussed further reading.

APPENDIX A

Vector Calculus

For a scalar function $C(\mathbf{x})$ that takes a vector argument $\mathbf{x} = [x_x, x_y, x_z]$, the gradient is computed as

$$\frac{\partial C(\mathbf{x})}{\partial \mathbf{x}} = \begin{bmatrix} \dfrac{\partial C(\mathbf{x})}{\partial x_x} \\[2ex] \dfrac{\partial C(\mathbf{x})}{\partial x_y} \\[2ex] \dfrac{\partial C(\mathbf{x})}{\partial x_z} \end{bmatrix}. \tag{A.1}$$

The derivative of a vector function $\mathbf{f}(\mathbf{x}) = [f_x(\mathbf{x}), f_y(\mathbf{x}), f_z(\mathbf{x})]$ with respect to the vector $\mathbf{x}$ is given by

$$\frac{\partial \mathbf{f}(\mathbf{x})}{\partial \mathbf{x}} = \begin{bmatrix} \dfrac{\partial f_x(\mathbf{x})}{\partial x_x} & \dfrac{\partial f_x(\mathbf{x})}{\partial x_y} & \dfrac{\partial f_x(\mathbf{x})}{\partial x_z} \\[2ex] \dfrac{\partial f_y(\mathbf{x})}{\partial x_x} & \dfrac{\partial f_y(\mathbf{x})}{\partial x_y} & \dfrac{\partial f_y(\mathbf{x})}{\partial x_z} \\[2ex] \dfrac{\partial f_z(\mathbf{x})}{\partial x_x} & \dfrac{\partial f_z(\mathbf{x})}{\partial x_y} & \dfrac{\partial f_z(\mathbf{x})}{\partial x_z} \end{bmatrix}. \tag{A.2}$$

DERIVATIVE CHAIN RULES

$$\frac{\partial}{\partial \mathbf{x}} (f(g(x))) = \frac{\partial f(u)}{\partial u} \frac{\partial g(x)}{\partial x}$$

$$\frac{\partial}{\partial \mathbf{x}} (f(\mathbf{x})g(\mathbf{x})) = \frac{\partial f(\mathbf{x})}{\partial \mathbf{x}} g(\mathbf{x}) + \frac{\partial g(\mathbf{x})}{\partial \mathbf{x}} f(\mathbf{x}) \qquad \text{(A.3)}$$

$$\frac{\partial}{\partial \mathbf{x}} \left(\frac{f(\mathbf{x})}{g(\mathbf{x})} \right) = \frac{\frac{\partial f(\mathbf{x})}{\partial \mathbf{x}} g(\mathbf{x}) - \frac{\partial g(\mathbf{x})}{\partial \mathbf{x}} f(\mathbf{x})}{g(\mathbf{x})^2}.$$

VECTOR EQUALITIES

$$\hat{\mathbf{x}} = \frac{\mathbf{x}}{||\mathbf{x}||}$$

$$||\mathbf{x}|| = \sqrt{\mathbf{x} \cdot \mathbf{x}} = \sqrt{x_x^2 + x_y^2 + x_z^2}$$

$$\frac{\partial ||\mathbf{x}||}{\partial \mathbf{x}} = \frac{\mathbf{x}^T}{||\mathbf{x}||} = \hat{\mathbf{x}}^T \qquad \text{(A.4)}$$

$$\frac{\partial \hat{\mathbf{x}}}{\partial \mathbf{x}} = \frac{\partial \left(\frac{\mathbf{x}}{||\mathbf{x}||} \right)}{\partial \mathbf{x}} = \frac{\mathbf{I}||\mathbf{x}|| - \mathbf{x}\hat{\mathbf{x}}^T}{||\mathbf{x}||^2} = \frac{\mathbf{I} - \hat{\mathbf{x}}\hat{\mathbf{x}}^T}{||\mathbf{x}||}.$$

Bibliography

T. Akenine-Möller, E. Haines, N. Hoffman, A. Pesce, M. Iwanicki, and S. Hillaire, *Real-time Rendering*, 4th ed., AK Peters/CRC Press, 2018. 86

D. Baraff and A. Witkin, Large steps in cloth simulation, *Proc. of the 25th Annual Conference on Computer Graphics and Interactive Techniques*, pp. 43–54, ACM, 1998. DOI: 10.1145/280814.280821. viii, 5, 19, 31, 41, 44, 57, 59, 73, 88, 92

D. Baraff, A. Witkin, and M. Kass, Untangling cloth, *ACM Transactions on Graphics (TOG)*, vol. 22, no. 3, pp. 862–870, 2003. DOI: 10.1145/882262.882357. 90

K. S. Bhat, C. D. Twigg, J. K. Hodgins, P. K. Khosla, Z. Popović, and S. M. Seitz, Estimating cloth simulation parameters from video, *Proc. of the ACM SIGGRAPH/Eurographics Symposium on Computer Animation*, pp. 37–51, 2003. 93

J. Bender, M. Müller, and M. Macklin, Position-based simulation methods in computer graphic, *EUROGRAPHICS Tutorials*, Eurographics Association, 10.2312/egt.20171034, 2017. DOI: 10.1111/cgf.12346. 91

M. Bergou, S. Mathu, M. Wardetzky, and E. Grinspun, TRACKS: Toward directable thin shells, *ACM Transactions on Graphics (TOG)*, vol. 26, no. 3, p. 50, 2007. DOI: 10.1145/1276377.1276439. 93

M. Botsch, L. Kobbelt, M. Pauly, P. Alliez, and B. Lévy, *Polygon Mesh Processing*, CRC Press, 2010. DOI: 10.1201/b10688. 12

S. Bouaziz, S. Martin, T. Liu, K. Ladislav, and M. Pauly, Projective dynamics: Fusing constraint projections for fast simulation, *ACM Transactions on Graphics (TOG)*, vol. 33, no. 4, p. 154, 2014. DOI: 10.1145/2601097.2601116. 91

D. E. Breen, D. H. House, and Ph. H. Getto, A physically-based particle model of woven cloth, *The Visual Computer*, vol. 8, no. 5–6, pp. 264–277, 1992. DOI: 10.1007/bf01897114. 5

D. E. Breen, D. H. House, and M. J. Wozny, Predicting the drape of woven cloth using interacting particles, *Proc. of the 21st Annual Conference on Computer Graphics and Interactive Techniques*, pp. 365–372, 1994. DOI: 10.1145/192161.192259. 57

R. Bridson, R. Fedkiw, and J. Anderson, Robust treatment of collisions, contact and friction for cloth animation, *ACM Transactions on Graphs*, vol. 21, no. 3, pp. 594–603, 2002. DOI: 10.1145/1198555.1198572. 89, 90, 93

R. Bridson, S. Marino, and R. Fedkiw, Simulation of clothing with folds and wrinkles, *Proc. of the ACM SIGGRAPH/Eurographics Symposium on Computer Animation*, pp. 28–36, 2003. DOI: 10.1145/1198555.1198573. 88, 92

M. Carignan, Y. Yang, N. Magnenat-Thalmann, and D. Thalmann, Dressing animated synthetic actors with complex deformable clothes, *Computer Graphics Proceedings, Annual Conference Series, ACM SIGGRAPH*, pp. 92–104, 1992. DOI: 10.1145/142920.134017. 5

K. Choi and H. Ko, Stable but responsive cloth, *ACM SIGGRAPH*, 2002. DOI: 10.1145/1198555.1198571. 44, 92

L. D. Cutler, R. Gershbein, X. C. Wang, C. Curtis, E. Maigret, L. Prasso, and P. Farson, An art-directed wrinkle system for CG character clothing, *Proc. of the ACM SIGGRAPH/Eurographics Symposium on Computer Animation*, pp. 117–125, 2005. DOI: 10.1145/1073368.1073384. 93

E. De Aguiar, L. Sigal, A. Treuille, and J. K. Hodgins, Stable spaces for real-time clothing, *ACM Transactions on Graphics (TOG)*, vol. 29, no. 4, pp. 106, 2010. DOI: 10.1145/1778765.1778843. 92

M. Desbrun, P. Schröder, and A. Barr, Interactive animation of structured deformable objects, *Graphics Interface*, vol. 99, no. 5, p. 10, 1999. 44

T. DeRose, M. Kass, and T. Truong, Subdivision surfaces in character animation. *Proc. of the 25th Annual Conference on Computer Graphics and Interactive Techniques*, ACM 1998. DOI: 10.1145/280814.280826. 93

D. Dinev, T. Liu, and L. Kavan, Stabilizing integrators for real-time physics, *ACM Transactions on Graphics (TOG)*, vol. 37, no. 1, p. 9, 2018. DOI: 10.1145/3153420. 46

B. Eberhardt, A. Weber, and W. Strasser, A fast, flexible, particle-system model for cloth draping, *IEEE Computer Graphics and Applications*, vol. 16, no. 5, pp. 52–59, 1996. DOI: 10.1109/38.536275. 45

B. Eberhardt, O. Etzmuß, and M. Hauth, Implicit-explicit schemes for fast animation with particle systems, *Computer Animation and Simulation*, pp. 137–151, 2000. DOI: 10.1007/978-3-7091-6344-3_11. 44

C. Ericson, *Real-time Collision Detection*, CRC Press, 2004. DOI: 10.1201/b14581. 86, 88

C. R. Feynman, Modeling the appearance of cloth, MSc. thesis, Department of Electrical Engineering and Computer Science, MIT, Cambridge, MA, 1986. 5

E. Grinspun and P. Schröder, Normal bounds for subdivision-surface interference detection, *Proc. of the Conference on Visualization EEE Computer Society*, pp. 333–340, 2001. DOI: 10.1109/visual.2001.964529. 93

E. Grinspun, A. Hirani, M. Desbrun, and P. Schröder, Discrete shells, *Proc. of the ACM SIG-GRAPH/Eurographics Symposium on Computer Animation*, pp. 62–67, 2003. 92

F. Hahn, B. Thomaszewski, S. Coros, R. W. Sumner, F. Cole, M. Meyer, T. DeRose, and M. Gross, Subspace clothing simulation using adaptive bases, *ACM Transactions on Graphics (TOG)*, vol. 33, no. 4, p. 105, 2014. DOI: 10.1145/2601097.2601160. 92

D. R. Haumann, Modeling the physical behaviour of flexible objects, *ACM SIGGRAPH Course Notes no. 17—Topics in Physically-based Modeling*, 1987. 5

M. Hauth, Numerical technique for cloth simulation, *SIGGRAPH Course*, no. 29, 2003. 46

D. House and D. Breen, *Cloth Modeling and Animation*, AK Peters/CRC Press, 2000. DOI: 10.1201/9781439863947. 5

H. Iben, M. Meyer, L. Petrovic, O. Soares, J. Anderson, and A. Witkin, Artistic simulation of curly hair, *Proc. of the 12th ACM SIGGRAPH/Eurographics Symposium on Computer Animation*, pp. 63–71, 2013. DOI: 10.1145/2485895.2485913. 28

J. M. Kaldor, D. James, and S. Marschner, Simulating knitted cloth at the yarn level, *ACM Transactions on Graphics (TOG)*, vol. 27, no. 3, p. 65, 2008. DOI: 10.1145/1360612.1360664. 92

D. Kim, W. Koh, R. Narain, K. Fatahalian, A. Treuille, and J. F. O'Brien, Near-exhaustive precomputation of secondary cloth effects, *ACM Transactions on Graphics (TOG)*, vol. 32, no. 4, p. 87, 2013. DOI: 10.1145/2461912.2462020. 92

W. Koh, R. Narain, and J. F. O'Brien, View-dependent adaptive cloth simulation, *Proc. of the ACM SIGGRAPH/Eurographics Symposium on Computer Animation*, pp. 1–8, 2014. 93

R. Kondo, T. Kanai, and K. Anjyo, Directable animation of elastic objects, *Proc. of the ACM SIGGRAPH/Eurographics Symposium on Computer Animation*, pp. 127–134, 2005. DOI: 10.1145/1073368.1073385. 76, 93

M. Li, A. Sheffer, N. Vining, and E. Grinspun, FoldSketch: Enriching garments with physically reproducible folds, *SIGGRAPH*, 2018. DOI: 10.1145/3197517.3201310. 92

S. Li, J. Huang, M. Desbrun, and X. Jin, Interactive elastic motion editing through space—time position constraints, *Computer Animation and Virtual Worlds*, vol. 24, no. 3, pp. 409–417, 2013. DOI: 10.1002/cav.1521. 76

L. Ling, *Aerodynamic Effects in Cloth Modeling and Animation*, D. H. House and D. E. Breen (Eds.), pp. 175–195, AK Peters, Ltd., Natick, MA, 2000. 14

T. Liu, A. Bargteil, J. F. O'Brien, and K. Ladislav, Fast simulation of mass-spring systems, *ACM Transactions on Graphics (TOG)*, vol. 32, no. 6, p. 214, 2013. DOI: 10.1145/2508363.2508406. viii, 47, 51

M. Macklin, M. Müller, and N. Chentanez, XPBD: Position-based simulation of compliant constrained dynamics, *Proc. of the 9th International Conference on Motion in Games*, pp. 49–54, 2016. DOI: 10.1145/2994258.2994272. 91

A. McNamara, A. Treuille, Z. Popović, and J. Stam, Fluid control using the adjoint method, *ACM Transactions on Graphics (TOG)*, vol. 23, no. 3, pp. 449–456, 2004. DOI: 10.1145/1015706.1015744. 76, 78

N. Metaaphanon, Y. Bando, B. Y. Chen, and T. Nishita, Simulation of tearing cloth with frayed edges. *Computer Graphics Forum*, vol. 28, no. 7, pp. 1837–1844, 2009. DOI: 10.1111/j.1467-8659.2009.01561.x. 28, 92

M. Müller, B. Heidelberger, M. Hennix, and J. Ratcliff, Position based dynamics, *Journal of Visual Communication and Image Representation*, vol. 18, no. 2, pp. 109–118, 2007. DOI: 10.1016/j.jvcir.2007.01.005. 91

J. Nocedal and S. Wright, Numerical optimization, *Springer Science*, vol. 35, pp. 67–68, 1999. DOI: 10.1007/b98874. 78, 82

T. Pfaff, R. Narain, J. M. de Joya, and J. F. O'Brien, Adaptive tearing and cracking of thin sheets, *ACM Transactions on Graphics*, vol. 33, no. 4, pp. 1–9, 2014. DOI: 10.1145/2601097.2601132. 92

D. Pritchard, Implementing Baraff and Witkin's cloth simulation, `http://davidpritchard.org/freecloth/docs/report.pdf`, 2006. 61, 71

X. Provot, Deformation constraints in a mass-spring model to describe rigid cloth behavior, *Graphics Interface*, Canadian Information Processing Society, 1995. 5

X. Provot, Collision and self-collision handling in cloth model dedicated to design garments, *Computer Animation and Simulation*, pp. 177–189, 1997. DOI: 10.1007/978-3-7091-6874-5_13. 90

S. C. Schvartzman, A. G. Perez, and M. A. Otaduy, Star-contours for efficient hierarchical self-collision detection, *ACM Transactions on Graphics (Proc. of ACM SIGGRAPH)*, vol. 29, no. 3, 2010. DOI: 10.1145/1833349.1778817. 90

A. Selle, M. Lentine, and R. Fedkiw, A mass spring model for hair simulation, *ACM Transactions on Graphics (TOG)*, vol. 27, no. 3, pp. 64, 2008. DOI: 10.1145/1360612.1360663. 28

J. Shewchuk, An introduction to the conjugate gradient method without the agonizing pain, 1994. 41

T. Stuyck, Natural media simulation and art-directable simulations for computer animation, KU Leuven Ph.D. thesis, 2017. 83

T. Stuyck and Ph. Dutré, Sculpting fluids: A new and intuitive approach to art-directable fluids, *ACM SIGGRAPH Posters*, p. 11, 2016. DOI: 10.1145/2945078.2945089. 83

T. Stuyck and Ph. Dutré, Model predictive control for robust art-directable fluids, *ACM SIG-GRAPH Posters*, p. 10, 2016. DOI: 10.1145/2945078.2945088. 76

N. Magnenat-Thalmann, F. Cordier, M. Keckeisen, S. Kimmerle, R. Klein, and J. Meseth, Simulation of clothes for real-time applications, *Eurographics, Tutorials 1: Simulation of Clothes for Real-time Applications*, 2004. 5

R. Tamstorf and E. Grinspun, Discrete bending forces and their Jacobians, *Graphical Models*, vol. 75, no. 6, pp. 362–370, 2013. DOI: 10.1016/j.gmod.2013.07.001. 73

D. Terzopoulos, J. Platt, A. Barr, and K. Fleischer, Elastically deformable models, *Computer Graphics Proceedings, Annual Conference Series. ACM SIGGRAPH*, pp. 205–214, 1987. DOI: 10.1145/37402.37427. 5

M. Teschner, B. Heidelberger, M. Muller, and M. Gross, A versatile and robust model for geometrically complex deformable solids, *Computer Graphics International*, pp. 312–319, 2004. DOI: 10.1109/cgi.2004.1309227. 29

P. Volino and N. Magnenat-Thalmann, Virtual clothing: Theory and practice, *Springer Science and Business Media*, 2000. DOI: 10.1007/978-3-642-57278-4. 5, 90

H. Wang, R. Ramamoorthi, and J. F. O'Brien, Data-driven elastic models for cloth: Modeling and measurement, *ACM Transactions on Graphics, Proceedings of ACM SIGGRAPH*, vol. 20, no. 4, pp. 71–82, 2011. DOI: 10.1145/2010324.1964966. 73

J. Weil, The synthesis of cloth objects, *SIGGRAPH Computer Graphics*, vol. 20, no. 4, pp. 49–54, 1986, DOI: 10.1145/15886.15891. 4

C. Wojtan, P. Mucha, and G. Turk, Keyframe control of complex particle systems using the adjoint method, *Proc. of the ACM SIGGRAPH/Eurographics Symposium on Computer Animation*, pp. 15–23, 2006. viii, 75, 78, 79, 80, 82

Author's Biography

TUUR STUYCK

Tuur Stuyck started working with computer graphics as early as 2006. He created several computer-animated short stories that were selected by international film festivals and won a number of prizes at national youth film festivals. He later became a jury member for the MakingMovies and Kunstbende youth film festivals. In addition to working on his own short movies during his high school and university years, Tuur also worked summer jobs at Cyborn Animation Studios in Antwerp, Belgium.

In 2008, he started his Bachelor's degree in engineering at KU Leuven, where he majored in Computer Science and minored in Electrical Engineering and graduated cum laude in 2011. In 2013, Tuur graduated magna cum laude with a Master's degree in Mathematical Engineering from KU Leuven. He obtained his Ph.D. in the Computer Graphics Research Group under the supervision of Prof. Dr. ir. Philip Dutré in 2017. During his Ph.D. research, he was awarded second place in the 2016 ACM SIGGRAPH research competition for his work on art directable simulations.

Tuur also collaborated with Adobe Research working on natural media simulation. Additionally, Stuyck interned twice at Pixar Animation Studios researching art-directed cloth simulations for feature film production. After graduating, he joined Pixar as a Postdoctoral Research Scientist. He currently works as a Postdoctoral Research Scientist at Facebook Reality Labs (previously known as Oculus Research).

Printed in the United States
by Baker & Taylor Publisher Services